T0205552

Advances in Intelligent Systems and Computing

Volume 1238

The series "Advances in Intelligent Systems and Computing" contains publications on theory, applications, and design methods of Intelligent Systems and Intelligent Computing. Virtually all disciplines such as engineering, natural sciences, computer and information science, ICT, economics, business, e-commerce, environment, healthcare, life science are covered. The list of topics spans all the areas of modern intelligent systems and computing such as: computational intelligence, soft computing including neural networks, fuzzy systems, evolutionary computing and the fusion of these paradigms, social intelligence, ambient intelligence, computational neuroscience, artificial life, virtual worlds and society, cognitive science and systems, Perception and Vision, DNA and immune based systems, self-organizing and adaptive systems, e-Learning and teaching, human-centered and human-centric computing, recommender systems, intelligent control, robotics and mechatronics including human-machine teaming, knowledge-based paradigms, learning paradigms, machine ethics, intelligent data analysis, knowledge management, intelligent agents, intelligent decision making and support, intelligent network security, trust management, interactive entertainment, Web intelligence and multimedia.

The publications within "Advances in Intelligent Systems and Computing" are primarily proceedings of important conferences, symposia and congresses. They cover significant recent developments in the field, both of a foundational and applicable character. An important characteristic feature of the series is the short publication time and world-wide distribution. This permits a rapid and broad dissemination of research results.

** **Indexing: The books of this series are submitted to ISI Proceedings, EI-Compendex, DBLP, SCOPUS, Google Scholar and Springerlink** **

More information about this series at http://www.springer.com/series/11156

Javier Prieto · António Pinto ·
Ashok Kumar Das · Stefano Ferretti
Editors

Blockchain and Applications

2nd International Congress

 Springer

Editors
Javier Prieto
BISITE Research Group
University of Salamanca
Salamanca, Spain

Ashok Kumar Das
Center for Security, Theory
and Algorithmic Research
IIIT
Hyderabad, Telangana, India

António Pinto
Politecnico do Porto and INESC TEC
Porto, Portugal

Stefano Ferretti
Department of Pure and Applied Sciences
University of Urbino
Urbino, Italy

ISSN 2194-5357   ISSN 2194-5365   (electronic)
Advances in Intelligent Systems and Computing
ISBN 978-3-030-52534-7   ISBN 978-3-030-52535-4   (eBook)
https://doi.org/10.1007/978-3-030-52535-4

This Springer imprint is published by the registered company Springer Nature Switzerland AG
The registered company address is: Gewerbestrasse 11, 6330 Cham, Switzerland

Preface

The 2nd International Conference on Blockchain and Applications 2020 (BLOCKCHAIN'20), held in the Heritage city of L'Aquila (Italy), has been a meeting point for both experienced and young researchers investigating in the areas of blockchain and artificial intelligence (AI). The conference has acted as a forum at which the attendees listened to lectures, and shared and discussed ideas, projects and advances associated with these technologies and their application domains. Within the scientific community, blockchain and AI are viewed as a promising combination that will transform the production and manufacturing industry, media, finance, insurance, e-government, etc. Nevertheless, there is no consensus with schemes or best practices that would specify how blockchain and AI should be used together. Combining blockchain mechanisms and artificial intelligence is still a particularly challenging task, and the BLOCKCHAIN'20 conference has been a milestone towards its achievement.

The BLOCKCHAIN'20 conference has been devoted to promoting the investigation of cutting-edge blockchain technology, exploring the latest blockchain- and AI-related ideas, innovations, guidelines, theories, models, technologies, applications and tools for the industry. Critical issues and challenges have been identified so that researchers and practitioners may address them in future research. The technical programme has been carefully designed to offer a fresh and balanced selection of advances and results in blockchain and AI, encouraging focus on novel and interdisciplinary topics.

The technical programme has been diverse and of high quality, and it focused on contributions to both well-established and evolving areas of research. More than 40 papers have been submitted to 20 from over 20 different countries (Canada, France, Germany, India, Ireland, Italy, Jordan, Luxembourg, Malaysia, Malta, Morocco, Netherlands, Oman, Portugal, Slovenia, Spain, Sweden, UAE and USA).

We would like to thank all the contributing authors, the members of the Programme Committee, the sponsors (IBM, Indra, EurAI, AEPIA, AFIA, APPIA and AIR Institute) and the Organizing Committee for their hard and highly valuable work. We thank the funding supporting with the project "Intelligent and sustainable mobility supported by multi-agent systems and edge computing" (Id. RTI2018-095390-B-C32); their work contributed to the success of the BLOCKCHAIN'20 event, and finally, we thank the Local Organization members and the Programme Committee members for their hard work, which was essential for the success of BLOCKCHAIN'20.

Javier Prieto
António Pinto
Ashok Kumar Das
Stefano Ferretti

Organization

General Chair

Javier Prieto Tejedor   University of Salamanca, Spain and AIR
Institute, Spain

Advisory Board

António Pinto   Instituo Politécnico do Porto, Portugal

Programme Committee Chairs

Ashok Kumar Das   IIIT Hyderabad, India
Abdelhakim Hafid   Université de Montréal, Canada

Local Chair

Stefano Ferretti   University of Bologna, Italy

Programme Committee

Amr Youssef   Concordia University, Canada
André Zúquete   University of Aveiro, Portugal
Andrea Omicini   Alma Mater Studiorum–Università di Bologna,
Italy
Anne Laurent   LIRMM - UM, France
Arnaud Castelltort   Montpellier University, France
Chhagan Lal   University of Padova, Italy
Daniel Jesus Munoz Guerra   University of Malaga, Spain
David Rosado   University of Castilla-La Mancha, Spain
Fengji Luo   The University of Sydney, Australia

Fernando De La Prieta	University of Salamanca, Spain
Francisco Luis Benítez Martínez	University of Granada, Spain
Georgios Samakovitis	University of Greenwich, UK
Giovanni Ciatto	University of Bologna, Italy
Hélder Gomes	Escola Superior de Tecnologia e Gestão de Águeda, Universidade de Aveiro, Portugal
Imtiaz Ahmad Akhtar	Higher Colleges of Technology, Qatar
João Paulo Magalhaes	ESTGF, Porto Polytechnic Institute, Portugal
Josep Lluis De La Rosa	EASY Innovation Center, UdG & RPI, Spain
Joshua Ellul	University of Malta, Malta
Kaiwen Zhang	École de technologie supérieure de Montréal, Canada
Kashif Zia	Sohar University, Oman
Luis Carlos Martínez	University of Salamanca, Spain
Manuel E. Correia	CRACS/INESC TEC; DCC/FCUP, Portugal
Marc Jansen	University of Applied Sciences Ruhr West, Germany
Marco Vitale	Foodchain Spa, Italy
Massimo Bartoletti	Dipartimento di Matematica e Informatica, Università degli Studi di Cagliari, Italy
Matthias Pohl	Otto-von-Guericke-Universität Magdeburg, Germany
Miguel Frade	Instituto Politécnico de Leiria, Portugal
Mirko Zichichi	Universidad Politécnica de Madrid, Spain
Mohamed Laarabi	Mohammadia School of Engineering Rabat, Morocco
Odelu Vanga	Birla Institute of Technology & Science (BITS), Pilani, Hyderabad Campus, India
Raja Jurdak	Commonwealth Scientific Industrial and Research Organization, UK
Ricardo Alonso	University of Salamanca, Spain
Ricardo Santos	ESTG/IPP, Portugal
Roberto Di Pietro	Hamad Bin Khalifa University - College of Science and Engineering, Saudi Arabia
Roberto Zunino	University of Trento, Italy
Roberto Casado-Vara	University of Salamanca, Spain
Rogério Reis	University of Porto, Portugal
Sami Albouq	Islamic University of Madinah, Saudi Arabia
Stefano Mariani	Università degli Studi di Modena e Reggio Emilia, Italy
Subhasis Thakur	National University of Ireland, Galway, Ireland
Vasilios Siris	Athens University of Economics and Business, Greece

Vicente Traver   Universitat Politècnica de València, Spain
Yuansong Qiao   Athlone Institute of Technology, Ireland

Organizing Committee

Juan M. Corchado Rodríguez   University of Salamanca, Spain
                                  AIR Institute, Spain
Javier Prieto Tejedor   University of Salamanca, Spain
                                  AIR Institute, Spain
Roberto Casado Vara   University of Salamanca, Spain
Fernando De la Prieta   University of Salamanca, Spain
Sara Rodríguez González   University of Salamanca, Spain
Pablo Chamoso Santos   University of Salamanca, Spain
Belén Pérez Lancho   University of Salamanca, Spain
Ana Belén Gil González   University of Salamanca, Spain
Ana De Luis Reboredo   University of Salamanca, Spain
Angélica González Arrieta   University of Salamanca, Spain
Emilio S. Corchado   University of Salamanca, Spain
  Rodríguez
Angel Luis Sánchez Lázaro   University of Salamanca, Spain
Alfonso González Briones   University Complutense of Madrid, Spain
Yeray Mezquita Martín   University of Salamanca, Spain
Enrique Goyenechea   University of Salamanca, Spain
                                  AIR Institute, Spain
Javier J. Martín Limorti   University of Salamanca, Spain
Alberto Rivas Camacho   University of Salamanca, Spain
Ines Sitton Candanedo   University of Salamanca, Spain
Elena Hernández Nieves   University of Salamanca, Spain
Beatriz Bellido   University of Salamanca, Spain
María Alonso   University of Salamanca, Spain
Diego Valdeolmillos   AIR Institute, Spain
Sergio Marquez   University of Salamanca, Spain
Jorge Herrera   University of Salamanca, Spain
Marta Plaza Hernández   University of Salamanca, Spain
Guillermo Hernández   AIR Institute, Spain
  González
Luis Carlos Martínez   University of Salamanca, Spain
  de Iturrate   AIR Institute, Spain
Ricardo S. Alonso Rincón   University of Salamanca, Spain
Javier Parra   University of Salamanca, Spain
Niloufar Shoeibi   University of Salamanca, Spain
Zakieh Alizadeh-Sani   University of Salamanca, Spain

Local Organizing Committee

Pierpaolo Vittorini	University of L'Aquila, Italy
Tania Di Mascio	University of L'Aquila, Italy
Giovanni De Gasperis	University of L'Aquila, Italy
Federica Caruso	University of L'Aquila, Italy
Alessandra Galassi	University of L'Aquila, Italy

BLOCKCHAIN'20 Sponsors

Sponsors   Organizers

Support from National Associations

Contents

BLOCKCHAIN-MainTrack

Sandboxes and Testnets as "Trading Zones" for Blockchain Governance

Denisa Reshef Kera[✉]

BISITE, University of Salamanca, Edificio I+D+i - C/Espejo s/n,
37007 Salamanca, Spain
denisa.kera@usal.es

Abstract. FinTech regulatory sandboxes and testnets use cases offer a hybrid model for integrating blockchain technologies with governance, connecting code with regulations, on-chain infrastructure with off-chain institutions. The hybrid models are an alternative to the reduction of governance to consensus mechanisms in the present libertarian but also anarcho-capitalist and communitarian blockchain projects. Inspired by the concepts of "innovation through dissonance" in the so-called "trading zones," we claim that the regulatory sandboxes can integrate all four regulatory forces (law, social norms, market, and technical infrastructure) rather than only two (FinTech insistence on markets and technology). This evaluation criterium for sandboxes was defined and tested with a simulated ledger (testnet) for exploring near-future scenarios of blockchain governance. In 2019, we conducted five workshops with 35 participants using templates of smart contracts to decide upon regulations of novel services that use satellite data to trigger automatic transactions (change of ownership). In the workshop and following questionnaire, the participants expressed need for a better integration of their natural language, regulations, and code without prioritizing any regulatory force or domain (market, culture, technology, or law), but supporting what we describe as a playful "regulation through dissonance."

Keywords: Blockchain · Governance · Regulatory sandbox

1 Introduction

Blockchain applications are often embraced, but also rejected for their ability to disrupt existing institutions and regulations in the financial services, land registries, and various industries. Bitcoin cryptocurrency, self-regulating and anonymous DAOs (Decentralized Autonomous Organizations), and smart contracts promise a more efficient, transparent, and decentralized governance reduced to algorithms. They claim to embody various libertarian [4], but also communitarian values and aspirations [1], such as "credible neutrality" [3] or "Ostrom's eight principles for commons stewardship" [11], but it remains unclear who decides on the values and how exact should be their mapping to the algorithms and code which are prone to changes.

Like any software, blockchain technologies suffer from security flaws and they need occasional maintenance. Any change in the code makes essential the coordination between the stakeholders, such as developers, miners, and users. Paradoxically, the

J. Prieto et al. (Eds.): BLOCKCHAIN 2020, AISC 1238, pp. 3–12, 2020.
https://doi.org/10.1007/978-3-030-52535-4_1

technology that is supposed to disrupt all governance has severe governance deficit when it comes to responding to the everyday challenges (common in every infrastructure) of maintenance, scaling or security flaws.

This lack of management of the actual software by its stakeholders leads to crises, and different fractions split to make their own version of the ledger and this weakens the original network. These so called "forks" of the mainnet (the main network of nodes that form the distributed ledger and the core functionality), but also testnets (simulated ledgers for testing of new applications) further erode the trust in the blockchain platforms.

There are many critiques of the governance by blockchain idea that expects algorithms or consensus mechanisms to mature and replace all existing institutions [6, 13]. What is often neglected in these discussions is the emergence of an alternative to these purist and reductionist views of governance by algorithms and code. It introduces a more hybrid model for convergence of blockchain technology with governance institutions, markets etc. via the regulatory sandboxes. Sandboxes defy the reductionist view of "governance-by-design" and introduce a more pragmatic model for adoption of blockchain technologies that can extended to other than FinTech domains.

2   Regulatory Sandboxes, Testnets, and Other Simulators

Regulatory sandboxes, but also testnets use cases, and simulators offer an alternative to the exaggerated social and political promises and threats of the blockchain technologies. They replace the discourse on disruption with actual experiments in contained and supervised environments that support stakeholder negotiations. Their goal is to integrate technology with governance that avoids the pitfalls of technocratic determinism and reductionism or equally restrictive dream of ex-ante regulations preventing any innovation in the name of "slow" governance.

The hybrid and alternative model of blockchain sandboxes was pioneered in the FinTech domain in 2015 by the UK Financial Conduct Authority as part of their innovation program (Innovate).[1] FCA's goal was to create a FinTech "ecosystem" that can negotiate and supervise the interactions between stakeholders and their interests. In the "sandbox," the innovators, existing financial institutions, but also government regulators negotiate and experiment with new services and combine various agendas: regulatory compliance, innovation, but also inclusivity and diversity.

The regulatory sandboxes simply extended the concept of a testing environment commonly used in software development and computer security to explore the interaction between emerging technology and society, regulations and code. In software development and security, a sandbox usually means a virtual server or other isolated and controlled environment, in which we can test how a piece of code interacts with a given operating system or various programs. In the case of a regulatory sandbox,

---

[1] FCA 'Regulatory Sandbox' https://www.fca.org.uk/firms/regulatory-sandbox.

multiple regulators set up an environment by "relaxing" the rules to "live-test" the emerging technologies on a limited sample of users.

The sandbox is usually set up by the provider of the platform (or regulations) to support better integration and symbiosis with something developed by "third parties" or private companies. The purpose is to mitigate the risks on all sides and anticipate the changes that will support the adoption without disruption and instability. The hybrid model of connecting regulations and code then offers some unique possibilities how to avoid the paradoxes of blockchain governance, but also the extremes of the "governance by design" [10] or regulatory moratoria.

3 CC License as the Origin of the "Regulatory Sandbox"

The search for pragmatic rather than reductionist solutions to the dilemmas of code and regulations started in the late 1990s with discussions on the "invisible regulation" by code (emerging technology platform or "architecture") in Lawrence Lessig's writing popularized as a "pathetic dot theory" [8].

According to Lessig, the individuals ("pathetic dots") are subjected to four regulatory forces (law, social norms, market, and architecture or technical infrastructure) that are not equally visible and negotiable. Lessig provides many examples of such regulatory (in)visibility to find a way to bridge the divide between code and law. Instead of making "better" code or more strict regulations, he proposed a hybrid initiative that connects these four forces by making them not only visible, but also negotiable to the stakeholders.

The 2001 CC license suite[2] is a simple tool and platform backed by an independent non-profit organization, Creative Commons (CC). It started as a proposal for regulation of digital content, but also a piece of reusable "code" included on various websites that offers a model (and license) for sharing and managing intellectual property. The CC license offered an alternative to the strict copyright model that was not working for digital content. It defined not only a new decentralized market for digital content without a "middleman" (copyright owners organizations), but also created new social norms around sharing online content by artists and creators.

In this sense, it impacted all four regulatory domains with a simple, but hybrid structure of regulation, code, and transformed social customs. The license empowered the citizens and stakeholders to engage with regulation, market, culture, and technology over simple icons and code that define what is a fair use for particular item, such an image, text or data. In this sense, it is a model for successful engagement with technological governance that is neither only about governance of technology nor only about reduction of governance to technology.

The regulatory sandboxes and examples of use cases based on the testnets extend these hybrid efforts for technological governance. They make the interaction between the four regulatory forces visible and negotiable to the stakeholders involved in the issue of blockchain adoption. The value of a sandbox for a blockchain governance

---

[2] *Creative Commons website* http://creativecommons.org/licenses/.

depends directly on how visible and negotiable it makes the four forces (regulation, market, technology, or culture) to the various stakeholders. Rather than reducing the impact to only one domain (better technology or market), sandboxes make use of the full spectrum of possibilities and connect them in novel ways.

4 Evaluation Criteria for Sandboxes and Hybrid Governance

Currently, the main criterium of regulatory sandbox success seems to be their ability to translate innovation to markets. This raises criticism and suspicion about their independence on the market and technology forces. To avoid this issue, we decided to define sandboxes more broadly as any institution, space, framework, method, or even a tool that makes the interaction between the four regulatory forces visible and open to experiments, discussion, and negotiation between different stakeholders.

Instead of insisting that code is an absolute law or that sacred laws should manage innovation by blocking certain code and making it compliant "by design," the purpose of a sandbox for blockchain governance is to support participatory experiments leading to integration and transformation of all four domains and forces of change.

Regulatory sandboxes, hybrid simulators, or certain uses of testnets have the potential to foster a symbiotic rather than antagonistic relation between code and regulations, platforms and institutions. They offer a practical alternative to the ex ante or ex post regulations coming too late to catch up with scandals and misuses of technology (Facebook, Cambridge Analytica, various Google services).

The few existing examples of regulatory sandboxes in the UK, but also Singapore and Australia, work mainly with FinTech projects. There are also many emerging sandboxes, such as the US-based sandbox set up by the Consumer Financial Protection Bureau (CFPB) for cryptocurrencies and blockchain technology in 2018 or 2019 Reserve Bank of South Africa sandbox. This makes it difficult to assess their value by independent sources, but according to the white papers and reports published about the first regulatory sandbox in UK, Financial Conduct Authority's (FCA) "Innovation" program enabled 11 blockchain and distributed ledger technology-related companies between 2015 and 2018.

Their narrow definition of success (market) raises doubts whether the purpose of sandboxes is actually to protect and improve the regulations. The emphasis on market can easily destabilize the existing public institutions in favor of the new businesses. To respond to these doubts, we are proposing a more inclusive criteria of success that relate to all four regulatory forces rather than two (technology and markets). The criteria include values such as visibility of the four forces shaping the future services, but also visibility and empowerment of the stakeholders by giving them a voice in the future development of the service. Regulatory sandbox success depends on how participatory and transparent the decision making, but also prototyping processes and how they can open they are to forces beyond the market.

Sandboxes should offer rich feedback on the type of issues, hopes, and fears the various stakeholders experience while engaging with the new service rather than only a quick recipe on how to enable adoption without too many regulations. The purpose of a sandbox for any emerging technology is to create trust, support diversity, and engagement rather than only innovation.

5 Sandboxes and Testnets as Trading Zones for "Regulation Through Dissonance"

To define the impact and evaluation criteria for such more open, antagonistic and hybrid forms of sandboxes for blockchain governance, we used the model described by anthropology and STS (Science, Technology, and society studies) scholarship as "trading zones" or "border crossings" [5, 9]. Trading zones are productive environments that offer a model for coordination and exchange of knowledge and resources between dissimilar and even antagonistic actors. We claim that regulatory sandboxes for emerging technologies, such as blockchain, should function as such trading zones between code and regulation rather than safe spaces for innovation without regulation.

Crucial role in the trading zones plays what Peter Galison described in his 1997 seminal work "Image and Logic: A Material Culture of Microphysics" [5] as a contact language. He defines it as a communication system or tool that enables participants to take "action" without changing their core "beliefs" or identity. Galison's classic example is the sociotechnical space (labs, but also tools) of the 20th-century particle physics. He describes how this "trading zone" served very diverse interests and agendas of the various types of scientists and other stakeholders to reform and innovate particle physics. The resulting particle physics achievement were not based on any unified theory, practice, value nor an institution, but on the "trade" between sometimes antagonistic groups of scientists and stakeholders' working together.

The exchange or trade between the antagonistic stakeholders was neither linear nor organized and controlled by one actor, goal or a tool. It was simply an open process, in which "pieces of devices, fragments of theories, and bits of language connect disparate groups of practitioners even when these practitioners disagree about their global significance" [5]. Particularly important for the success of such open and hybrid space and process was the possibility of reuse ("cannibalization") where the successful tools or ideas evolve rather than simply comply to some ideals: "Experimenters like to call their extractive moves "cannibalizing" a device. Televisions, bombs, computers, radios, all are taken apart, rearranged, and welded into the tools of the physicist. And the process can be inverted: instrumentation from physics becomes medical instruments, biological probes, and communication apparatus. Geiger-Muller counters were cannibalized to make the first electronic logic units for a computer, but pieces of the computer were soon stripped out for use in particle detectors" [5].

The model of a trading zone with a contact language that enables "cannibalization" of tools or ideas supports what is also described as an "innovation through dissonance" [12]. We extended this to a "regulation through dissonance" in the case of the blockchain sandboxes and testnets where different stakeholders work in an ad hoc integration of governance issues with new blockchain services.

Regulatory sandboxes and testnets are spaces that enable similar "cannibalization" and reuse of both regulation and code, institutions and consensus mechanisms, to form a hybrid interface, contact language or device between them. The hybrid environment supports actions and experiments with code and regulations by simply engaging various stakeholders that do not share the same goals and beliefs. Ideally, they create an opportunity or a tool that transforms all four agendas (new market, service, regulation and technology) without having one value or even main goal.

How does such "regulation through dissonance" in the hybrid sandboxes and trading zones actually look like? To illustrate this, we will use an example from our recent practice and explain the hybrid exchanges between regulation and code as such ad hoc process. These improvisations that are open to various stakeholders and all four regulatory forces support the principle of good governance rather than concrete regulation or technology.

We are starting to see more examples of such strategy, such as the CLAUDIA project, a testnet use case with a simple offchain reporting mechanism (compliance desk) serving DAO projects in Spain[3]. Our main example in this article is the blockchain simulation game, Lithopia (Lithopy), in which participants defined different ideas on how to regulate an imaginary smart village that uses smart contracts combining satellite and blockchain services.

6 Lithopia: Sandbox for Extreme Algorithmic Life

Lithopia started as a near future simulation game used in workshops exploring anticipatory governance of future blockchain infrastructure that use satellite and drone data to trigger smart contracts [7]. Some 35 participants took part in the five workshops in 2019 to experience "life" in the imaginary "smart village" of Lithopia under one of the 12 stakeholder roles.

The workshop would typically run for 4 h, during which participants would learn the basics of blockchain infrastructure and application design (Hyperledger Composer and Fabric), but also the type of regulatory possibilities in emerging technology (regulation through communication, audits, certificates, law, market, industry).

Participants would explore the interface for interacting with the blockchain infrastructure in the form of a dashboard (Node RED environment) and use templates of four smart contracts to deliberate on possible regulations of the blockchain and satellite services. We gathered data in a form of observations from the workshops, but also from the questionnaire (only 18 out of 35 participants left some feedback) which we are discussing in detail in a forthcoming paper (under review).

For the purpose of discussing a hybrid model of sandbox regulation, we will concentrate here on one important experience from the first two workshops where we realized that most participants struggled to identify with the assigned stakeholder role and the choices of regulatory strategies in relation to the code.

---

[3] https://github.com/kunfud/CLAUDIA.

Participants had strong personal opinions and stakes in the issues of automation and surveillance explored in the Lithopia smart contracts scenario (Hyperledger Fabric ledger with satellite services and various types and tools of regulations and self-regulation) and felt that they miss tools to express their concerns.

Several participants explicitly mentioned that they feel frustrated that their opinions, experiences, and knowledge do not really count when expressed in natural language without being translating into the language of code or regulation. They complained of being left out from the decision-making processes about the common future that is "outsourced to Silicon Valley code or EU commission elites' regulations" (example of descriptions used during the discussions in the second workshop with 9 participants).

One third of the participants from all five workshops refused the programming part as something they did not feel confident enough of doing, but also trusting it. In the second workshop, one participant explicitly asked for a better "translation" of the code into the natural language over an interface.

Rather than trying to find a model for regulation and code of Lithopia services in the workshop, participants expected Lithopia to serve as such an interface that allows more direct translation between natural languages, code, and regulations (without a human programmer as an intermediary). Two participants also expressed a similar requirement to hide the actual code of Lithopia and offer only the "logic" of the smart contracts expressed in a natural language.

Based on this feedback, in the last three workshops (August, October, December) we defined Lithopia to be such an interface and translation (sandbox) or "contact language" and "trading zone" between the different stakeholders. We started by confronting the participants with the limits of Hyperledger Composer and Fabric blockchain platform that commodifies all human interactions into four categories or entities used in the programming of the smart contracts (creation of an open API): assets, participants, transactions, and events. These four elements of what is called a "business network" (describing the organization or community) define the type of values saved and exchanged on the ledger.

To probe the idea of a "contact language" simplifying both code and regulations to enable more interaction we emphasized that the JavaScript logic of Lithopia smart contracts is readable and that the four categories of entities in the Lithopia business network can be freely interpreted. We probed how the extreme commodification ingrained in the basic architecture of the blockchain platform can be appropriated for very different agenda: the contracts in Lithopia define ownership of an asset only by the name of the owner without quantifying its value in currency.

The change of ownership is not about buying and selling with money but covering a space and creating a pixel of data for the satellite (10×10 m of red cloth). This "hack" of the Hyperledger Fabric platform introduces a possibility or stewardship and genealogy rather than ownership, and we intentionally left unclear if this is an act of vandalism, activism, or some land art type of intervention,

Most participants reacted to this "hack" with even stronger call for a service that would allow translation and management of the infrastructure and the technology via the natural language rather than specialized knowledge of syntax and code. They felt that the power of the platform is overwhelming and creates an asymmetric control.

Only few participants (5) in the last three workshops included a call for improving the coding literacy or some other unspecified support of understanding the technological infrastructure and the actual algorithms (more important for DAOs and public blockchains).

7   Lithopia Sandbox Challenges, Limits and Possibilities

The frustration with the asymmetry of power between the natural language of the different stakeholders, programming (code), and the bureaucratic system of regulations, certification processes etc. inspired the idea of a sandbox environment for "regulation through dissonance". The extreme scenarios that show how infrastructure can be tweaked to serve different purposes enabled the stakeholders discuss also issues of culture and politics rather than reducing everything to safe code or strict regulation.

In the three workshops that followed, we made the stakeholder roles optional and concentrated only on one of the smart contracts, in which we introduced a bias to make the discussions on code and regulations more focused and antagonistic. With this we tested what interventions make such sandbox comfortable enough for participants with various skills and agenda to act. In a "trading zone" participants could preserve their identity, agenda and even language to influence both regulation and code of the blockchain and satellite services in Lithopia.

Rather than deciding on the type of onchain and offchain regulation and modes or anticipatory governance, the Lithopia challenge transformed into a following question: How to enable a contact language between the code and regulation? How to "humanize" the exchanges with code and regulation that would make engagement between different stakeholders possible? Should there be an interface with explicit rules how code relates to regulations and natural language or we should accept the conflict and still insist on the interaction between different stakeholders? How to create an environment similar to the CC platform where we can test new model connecting the market with everyday culture and habits, but also regulations in the case of blockchain platforms?

There are several efforts to map our natural languages to law and algorithms that could enable a design that hides the complexities of regulations but also code and opens it to engagement via natural language (Project Cohubicol[4]). In the case of Lithopia workshops, we realized that it is not a matter of an explicit interface nor exact translations that serves as a contact language to change the code and rules of the satellite smart contracts. The impossible transactions between natural language, code and regulations showed that what matters is the actual engagement as a democratic and historical process of negotiations and conflict rather than forcing everyone to meet one goal (better regulation, better code or market adoption).

The sandbox type of experience like Lithopia workshop can help the citizens to decide upon a collective future outside of decontextualized game theory concepts of

---

[4] Cohubicol "new legal hermeneutics" from textual to computational law https://www.cohubicol.com/about.

various consensus mechanisms (governance by blockchain) or proposals to crowd-source data and attitudes of citizens over technological platforms [2]. Lithopia as a hybrid sandbox for experiencing the extremes of algorithmic governance and life shows also the limits of such engagements. The experience of the asymmetry of power and decision making goes beyond issues of communication or knowledge (learning code for example or having the ideal translator).

Instead of defining anticipatory governance (our starting point), we settled on more modest, but also ad hoc goals of connecting code with regulations as a time-limited experience of a "good enough" solutions in a particular situation. The biased smart contract that prevented a sale of property to anyone who was Czech proved to be an ideal "contact language" case where majority of participants (80% based on the observations and written feedback) preferred more principle-based rather than rule-based governance. They emphasized the importance of auditing codes and data based on some public-private partnerships in the future governance infrastructure (approximately 70%).

Rather than a particular rule, code, tool or regulation, there was a strong call for hybrid forms of auditing the onchain and offchain exchanges. Participants shared various ideas on how auditing of smart contracts and future platforms should look like raging from emphasis on independent agency "reading the code" and cooperating with the programmers to more technological, testnet based services for testing and communicating such future services to everyone (so everyone can see and test the code before it is implemented).

The participants that emphasized the importance of aspirations and principles in regulating future blockchain services insisted that they need to be open to interpretation and change through the historical and contextual processes of political or stakeholder deliberation and engagement (over natural language) rather than being hard-coded or over-regulated by experts.

This is a strong reason why the sandbox model is actually an ideal environment for designing and deciding together upon the blockchain future infrastructure and algorithmic governance: it enables a playful interaction between natural language, regulations and code without prioritizing any regulatory force or domain (market, culture, technology or law).

8 Summary

The function of a hybrid sandbox for blockchain governance is to serve as a trading zone for various stakeholders to make decisions that influence their common future. Stakeholders interact freely without a unified theory, technology or even policy goal over different parts of the code (architecture, technology) and regulations (governance). The purpose of interacting and mixing code and regulations is to supports the different stakeholders in working together in ad hoc manner in a process we describe as "regulation through dissonance" enabled by the sandboxes for participatory engagement in deliberation and prototyping of blockchain futures. In the case of our Lithopia example this resulted in strong emphasis on various forms of auditing of the code while implementing and designing future services. The hybrid sandbox for blockchain governance creates a medium of exchange that supports "action" while not insisting on any common goals. It supports exchanges, experimentation and negotiation between stakeholders that impact all four domains of regulation (market, culture, technology and governance) rather than only one. Success and value of a sandbox is then measured according to how many different (even antagonistic) stakeholders with various agendas, languages, ontology and interests it can support and how it impacts all four domains of regulation.

Acknowledgements. The research is supported by Horizon 2020 Marie Curie Individual Fellowship.

References

1. Atzori, M., Ulieru, M.: Architecting the eSociety on blockchain: a provocation to human nature. SSRN Electron. J. (2017)
2. Awad, E., Dsouza, S., Kim, R., et al.: The moral machine experiment. Nat. **563**(7729), 59–64 (2018)
3. Buterin, V.: Credible neutrality as a guiding principle (2020)
4. Flood, J., Robb, L.: Trust, anarcho-capitalism blockchain and initial coin offerings. SSRN Electronic J. (2017)
5. Galison, P.: Image and Logic : A Material Culture of Microphysics. University of Chicago Press, Chicago (1997)
6. Hassan, S., De Filippi, P.: The expansion of algorithmic governance: from code is law to law is code. Field Actions Sci. Rep., Special Issue **17**, 88–90 (2017)
7. Kera, D.R.: Anticipatory policy as a design challenge: experiments with stakeholders engagement in blockchain and distributed ledger technologies (BDLTS). In: Prieto, J., Das, A., Ferretti, S., Pinto, A., Corchado, J. (eds.) Blockchain and Applications, BLOCKCHAIN 2019. Advances in Intelligent Systems and Computing, pp. 87–92. Springer, Cham (2020)
8. Lessig, L.: The new Chicago school. J. Legal Stud. **27**(S2), 661–691 (1998)
9. Lewis, S.C., Usher, N.: Trading zones, boundary objects, and the pursuit of news innovation: a case study of journalists and programmers. Convergence **22**, 543–560 (2016)
10. Mulligan, D.K., Bamberger, K.A.: Saving governance-by-design. Calif. Law Rev. **106**(3), 697–784 (2018). https://doi.org/10.15779/Z38QN5ZB5H. Accessed 21 July 2019
11. Rozas, D., Tenorio-Fornés, A., Díaz-Molina, S., Hassan, S.: When ostrom meets blockchain: exploring the potentials of blockchain for commons governance. SSRN Electron. J. (2018)
12. Stark, D., Beunza, D., Girard, M., Lukacs, J.: The Sense of Dissonance : Accounts of Worth in Economic Life. Princeton University Press, Princeton (2011)
13. Trump, B.D., Wells, E., Trump, J., Linkov, I.: Cryptocurrency: governance for what was meant to be ungovernable. Environ. Syst. Decis. **38**(3), 426–430 (2018)

Functional Differences of Neo
and Ethereum as Smart Contract
Platforms

Marco Bareis, Monika di Angelo$^{(\boxtimes)}$, and Gernot Salzer

Faculty of Informatics, TU Wien, Favoritenstr. 9-11, 1040 Vienna, Austria
{marco.bareis,monika.di.angelo,gernot.salzer}@tuwien.ac.at

Abstract. While Ethereum is currently the most popular smart contract platform, there are interesting alternatives like Neo. However, Neo only rarely appears in scientific literature. Moreover, comparisons of smart contract platforms hardly employ a structured approach, but rather apply different criteria to each platform.

This work performs an in-depth comparison between Ethereum and Neo in a structured manner. We derive a catalogue of criteria from related work and use it to identify differences and similarities worthy of discussion. We show how Ethereum and Neo differ in key aspects, ranging from the general goal to technical issues like the execution and fee model and practical aspects like the maturity of its features and documentation.

Keywords: Evaluation · Criteria · Smart contract · Platform

1 Introduction

While Ethereum has been widely used as a platform for smart contracts (SCs), alternatives keep appearing. When a company decides on a SC platform for its business case, several aspects have to be considered. Since blockchain interoperability is still an issue, migrating applications from one platform to another can be expensive and time-consuming. Furthermore, when targeting the Asian market, the stance of China on this technology is an important factor.

One promising alternative to Ethereum is Neo, sometimes referred to as Chinese Ethereum, which at first glance indeed appears to be quite similar to Ethereum. An immediate difference emerges when looking for documentation. While documents about Ethereum are abundant and readily available, detailed information about Neo is sparse. Regarding comparisons, technical reviews tend to list the features of platforms individually instead of using a structured list of criteria as common reference. In particular, there is no rigorous comparison of Ethereum and Neo in terms of functionality and mechanics.

This paper is a condensed version of the comparison described in [2].

© The Editor(s) (if applicable) and The Author(s), under exclusive license
to Springer Nature Switzerland AG 2020
J. Prieto et al. (Eds.): BLOCKCHAIN 2020, AISC 1238, pp. 13–23, 2020.
https://doi.org/10.1007/978-3-030-52535-4_2

Contribution. This work elaborates in a structured manner the differences between Ethereum and Neo as SC platforms. Our contribution consists in (i) a catalogue of criteria relevant for the comparison of platforms, (ii) a comparative evaluation of Ethereum and Neo as SC platforms based on these criteria, and (iii) a discussion about the effects of the differences on the development of SCs.

Benefits and Audience. The catalogue provides a structured scheme for comparing SC platforms, which may serve as the basis for further well-founded comparative studies. Our work provides a guideline for companies and developers that have to choose between Ethereum and Neo.

Roadmap. Section 2 presents the criteria used for the comparison in Sect. 3. We discuss the differences in Sect. 4 and conclude in Sect. 5.

2   Catalogue of Criteria

After reviewing related work, we will specify the criteria for our comparison. They are highlighted in bold face and grouped into the categories *project, blockchain, platform,* and *operation.* We employed the following five-step approach to arrive at criteria relevant for comparing the functionality of SC platforms.

1. Collect descriptions and comparisons of blockchains or SC platforms,
2. extract terms used as criteria,
3. remove criteria not relevant for the functionality of SC platforms,
4. add criteria for features specific to Neo to avoid bias, and
5. merge similar terms.

2.1   Related Work

Early comparisons like [3,4] concentrate on Bitcoin and Ethereum. The survey [17] aims at breadth and describes 17 SC platforms, including Ethereum and Neo. A more recent one, [31], compares five platforms, among them Ethereum. Other publications focus on particular aspects: [24] compares the languages Solidity, Pact and Liquidity, while [30] is devoted to blockchain-based applications in general, highlighting Bitcoin, Ethereum and Hyperledger Fabric as SC platforms. [19] discusses scalability, interoperability and sustainability in general without referring to specific blockchains, while [1] compares consensus algorithms. [5] interviewed the developers of numerous blockchain projects on a broad variety of platforms. We used all mentioned publications, and some more, to develop our catalogue of critera. Moreover, we scanned [20] for Neo-specific criteria, as none of the papers but one relates to Neo.

2.2   Project

Objectives of a platform include specific use cases or application domains as well as general goals like decentralization. It also includes the possibility of integrating third-parties or government bodies. **Maturity** mainly refers to three aspects: (i) the platform is sufficiently mature for hosting applications in productive use [3], (ii) quality of documentation and (iii) status of specification. Developers of DAPPs need reliable information on the chosen platform and its mechanics. **Origin and Organization.** The country of origin and the organization behind a platform are relevant factors, when having to deal with legal regulations. **Governance** determines who can propose changes and who finally decides which changes will be implemented. Blockchain governance can be achieved by a variety of rules that are usually classified as on-chain or off-chain.

2.3   Blockchain

Chain and Dependencies includes the availability of a live mainnet and testnets as well as "whether the platform has its own blockchain, or if it just piggybacks on an already existing one" [3]. **Deployment Types** commonly divide networks into public, private and permissioned. To qualify as public (or permissionless), a chain must be accessible without special permission by anyone following the respective protocol [4]. **Consensus Protocol** is the most commonly mentioned criterion. It strongly influences key parameters like transaction volume, energy consumption and operation costs. It is crucial to the operation of a blockchain as it ensures reliability in a network of unreliable nodes; it guarantees the integrity and consistency of the blockchain.

2.4   Platform

Language Support means the availability of languages for writing SCs and the maturity of compilers and corresponding documentation. **Community** refers to the size and activity of the group of developers that is usually the first address to ask for help. **Execution Model.** The underlying mechanics need to guarantee three essential properties of SCs: determinism, isolation and termination [16]. Determinism is usually ensured by simply not offering non-deterministic functions or data sources and limiting or prohibiting dynamic calls. Isolation means that the executed SC is sandboxed, so that it cannot influence other contracts or the system itself. Finally, termination in Turing-complete SC platforms is usually achieved by limiting fees, the number of computation steps or time. **Interoperability** between SC platforms is achieved either when two or more platforms agree on a trusted third-party to transfer information and digital assets, or when the platforms share information directly with each other leveraging trust generated by SCs [19]. **Identity Management.** One of the key characteristics of existing blockchain platforms is pseudo-anonymity provided by addresses not linked to real identities. However, some applications require accountability and transparency in terms of user identity and therefore a secure association with real-life identities. **Tool Support** is crucial for an effective development process [24]. As most current tools are not tuned for SC development, the need for specialized tools like customized IDEs, debuggers and testing tools arises [5]. **Application Standards** improve interoperability, re-usability and security.

2.5   Operation

Block Time is the time between the creation of two subsequent blocks. **Block Confirmation Time** is the latency between submission and confirmation of a block. **Throughput** is the number of transactions processed per second. **Execution Costs.** Platforms employ a fee model to compensate nodes for the resources (computation and storage) used by SCs. Furthermore, fees prevent applications from consuming too many resources of a network.

3   Comparison

Based on the criteria from Sect. 2, we evaluate the available documents about the two platforms Ethereum and Neo. Some aspects need verification using actual SCs on the respective test chains. We also checked the Neo website and forum in Chinese with Google Translate to make sure nothing essential is overlooked.

3.1   Project

Objectives. Ethereum aims at a featureless generic platform that does not censor anything, supports the creation of decentralized applications or organizations [28,30], and is not dependent on a single country or its government.

Neo on the other side sets the 'smart economy' as its primary goal, and thus intends to incorporate features for identity management and cross-chain compatibility [20]. Smart economies need to work with government bodies, which Neo acknowledges by incorporating digital identity standards [25]. However, most of the features for supporting the smart economy are not yet finished.

Maturity. Ethereum has been live since July 2015, and provides a main net and test nets. Although not in its final form and subject to changes, it is the most used platform with thousands of DAPPs. Above all, Ethereum is well documented [12,28] has a formal description [29] and resources (like [12,13]).

Neo also provides a production network. However, it is not as widely used as Ethereum and most of the key features are still missing. Moreover, the platform is poorly documented. The whitepaper [20] is a mix of features of the current version (Neo 2.0) and the future version (Neo 3.0) and contains contradictory information. Furthermore, as of January 2020, Neo's yellow paper [7], that aims to define the technology formally, is mainly blank except for one section.

Origin and Organization. The development of Ethereum was funded using a crowdsale in 2015. Of this, 12 million Ether were retained for the Ethereum Foundation [11] with the mission to "promote and support Ethereum platform and base layer research, development and education to bring decentralized protocols and tools to the world that empower developers to produce next generation decentralized applications (DAPPs)" [13]. As such, they sign responsible for the development of the Ethereum client *Geth*, the SC scripting language Solidity and widely used development tools like Remix.

The Ethereum ecosystems consists of many different companies that participate in the development of Ethereum, infrastructure projects and DAPPs. Two organization with a high impact on Ethereum are ConsenSys [8] and the Enterprise Ethereum Alliance [10]. ConsenSys acts as an incubator for Ethereum related projects like MetaMask and facilitates communication and knowledge transfer between developers while allowing them to work autonomously [15]. Moreover, it acts as a venture capital company by financially supporting projects and provides services for companies who intend to incorporate Ethereum or create private Ethereum networks [15]. The importance of ConsenSys to Ethereum goes well beyond providing infrastructure tools. Because it ties to decentralized projects and companies implementing Ethereum, ConsenSys strongly influences the Ethereum project itself. Furthermore, it maintains close relationships to government bodies (like the European Commission).

Antshares (now Neo 1.0) was started by CEO Da Hongfei and CTO Erik Zhang, who already founded the blockchain company OnChain [27]. A crowdsale in 2015 raised funds to develop the platform. Antshares was rebranded to Neo in 2017 with a focus on the 'smart economy' [20]. The development of the Neo protocol is steered by the Neo Foundation, a Chinese non-profit organization, in which Hongfei and Zhang have executive authority [21]. In 2018, the companies Neo Global Development (NGD) and Neo Global Capital (NGC) were founded. NGD is a sub organization of the Neo Foundation and focuses on research & development, marketing, and community development [21]. NGC is a Singapore-based organization licensed for fund management.

The company most associated with Neo is OnChain. Da Hongfei even needed to clarify publicly that Neo and OnChain are in fact separate companies [23]. OnChain has its own blockchain product called Distributed Network Architecture (DNA), which helps other companies to set up blockchains. DNA is highly similar to Neo and therefore profits from its ongoing development. Da Hongfei even mentions that interoperability between Neo and DNA-based chains will be possible in the future [23]. OnChain closely cooperates with the Chinese government [26] with unclear impact.

Governance. Ethereum uses an off-chain governance process, which means that the rules are not encoded in the platform but applied on a social level. The governance process is based on the Ethereum Improvement Proposals (EIPs). EIPs are design documents that either present information to the community or describe a new feature of Ethereum or its surrounding processes. Although the process involves the Ethereum community, the core developers eventually determine which changes are implemented in the core protocol.

Neo utilizes off-chain and on-chain governance. Neo provides two types of native tokens: NEO and NeoGas. The indivisible token NEO represents the right to manage the network and participate in the on-chain governance process. Neither the whitepaper nor the yellow paper specify the concrete network parameters, on which the token holders can vote [20]. New ideas can be discussed in Neo Enhancement Proposals (NEP) [22], which work similar to EIPs in Ethereum. The final decision on proposals for the protocol is made by Hongfei and Zhang [9].

3.2   Blockchain Properties

Regarding the blockchain, Ethereum and Neo mainly differ in the **consensus protocol**. Ethereum uses Proof-of-Work (PoW) – with sustainability and scalability issues. With Ethereum 2.0, the switch to Proof-of-Stake (PoS) is planned, where anybody (with sufficient stake) can propose and validate blocks. Variants of PoS with incentives to facilitate decentralization are being discussed [6].

Neo puts a focus on scalability at the cost of a low degree of decentralization by using the Delegated Byzantine Fault Tolerant (dBFT) algorithm [20]. In fact, seven bookkeeping nodes propose and validate new blocks, of which cureently five (to six) nodes are under the control of the Neo Foundation (the two-third majority needed to accept or reject blocks on their own). All other nodes can indirectly influence the block creation by voting for one of the bookkeeping nodes. The founders of Neo intend to change this with Neo 3.0 [9].

3.3   Platform and Development

Language Support. Ethereum currently supports two programing languages actively in use, Solidity and Vyper. With Ethereum 2.0, plans are to add the Ethereum flavored WebAssembly (eWASM) as a second assembly platform [8] that inherits a wider range of languages and tools.

Neo supports common languages like Java or C# [20]. Being able to use the same programing languages for on- and off-chain parts eases development. However, some of the languages listed are not fully supported yet [20]. Moreover, documentation for the compilers (and their status) is lacking.

Community. Ethereum has an active community that provides help in many cases. Neo's community is much smaller and partly interacts in Chinese.

Execution Model. Both platforms execute SCs in a stack-based virtual machine that only offers deterministic functions. Isolation is achieved by sandboxing, while termination is guaranteed through a fee model (gas).

Platform Interoperability and Identity Management. Ethereum hosts DAPPs for both, with uPort being the most popular DAPP for identity management. Neo hosts a DAPP for identity management from the telecom provider Swisscom. It is planned to integrate both functionalities into the platform.

Tool Support. For Ethereum, ConsensSys and the large community provide tools, like the specialized IDE Remix, IDE add-ons and numerous further tools. Neo focuses on plug-ins for existing environments.

Application Standards. Ethereum offers a wide range of ERCs for tokens, identity management, proxy contracts and more. Neo offers a small number of application standards focusing on tokens.

3.4   Operation

Both platforms have a **block time** of around 15 s, but this is likely to change in future versions.

The **block confirmation time** for Ethereum is around six minutes, while the consensus protocol of Neo provides instant finality.

Ethereum achieves a **throughput** near 10 transactions per second, while Neo's promise of 10 000 transactions per second has not yet been verified.

Execution costs differ significantly. In Ethereum, everyone can deploy a contract for a few US Dollars. In contrast to that, in Neo one has to spend several thousand US Dollars per SC deployment. Thus, Neo is used by well-funded projects only. SC users encounter high fees as well. Neo's approach to compensate the high costs with grants further diminishes its decentralization.

3.5   Summary

Although both platforms look similar on a superficial level, the structured comparison revealed noteworthy differences. In Table 1, we marked those criteria in bold face where the platforms show notable differences.

4   Discussion

Both Ethereum and Neo are blockchain-based smart contract platforms featuring a Turing-complete virtual machine. This allows them to support all kinds of smart contract applications, and indeed both platforms host live applications.

The preference for a platform also depends on the market your application targets. If China is a main target for you DAPP, Neo might be a good choice. Furthermore, OnChain already creates blockchains for the Chinese government and Chinese companies. Those chains will eventually become compatible with Neo. Therefore, an entire blockchain ecosystem might emerge with Neo being the backbone of OnChain's blockchain concept. The plan is that Neo provides a public chain while OnChain provides private chains for enterprises, and the ultimate goal is to link both worlds together [27]. If the plan succeeds, Neo is likely to play a major role in the Chinese blockchain market. For developers focusing on the Chinese or Asian market, this is a strong argument in favor of Neo. China's interest in Neo may go hand in hand with stricter rules and higher demands. For example, when the Chinese government banned western social media platforms, they encouraged its population to use WeChat (developed by Tencent) [18]. Tencent on the other hand had to employ people close to the communist party. At least in theory, Neo currently outperforms Ethereum regarding scalability and language support. On the other hand, Ethereum is an established platform with an active community and quite likely will improve its performance and language support. Then again, Neo may become a globally successful blockchain as envisioned by its CEO Hongfei. The cooperation with Swisscom is a noteworthy step in that direction.

Ethereum's PoW results in a high power consumption [14] that will change
with the planned switch to PoS. Neo's power consumption is low already due to
its consensus mechanism and the small network consisting of only seven validating nodes and a comparably small total number of participating nodes.

Table 1. Summarized comparison of Ethereum and Neo

Criterion	Ethereum	Neo
Objectives	General purpose	Smart economy
Maturity	Live applications plentiful documentation formally specified VM	Live applications varying documentation no formal specification
Origin, organization	Ethereum foundation Switzerland company ConsenSys fairly independent	Neo foundation China company OnChain links to Chinese government
Governance	Off-chain core developers EIP	Off- and on-chain founders NEP
Chain	Mainnet and testnets	Mainnet and testnets
Deployment types	Permissionless	Permissionless
Consensus protocol	PoW (PoS), decentralized	dBFT, 7 validators
Language support	Domain specific languages, mature compiler and docs	Common general purpose Languages, almost mature compiler and docs
Community	Large and highly active	Small and active
Execution model	Deterministic, sand-boxed, resource pricing	Deterministic, sand-boxed, resource pricing
Platform interoperability	Not planned as built-in by means of SCs	Planned as built-in by means of SCs
Identity management	Not planned as built-in by means of SCs	Planned as built-in by means of SCs
Tool support	Plentiful in various stages of maturity IDE and add-ons	Inherited from common languages plug-ins
Application standards	ERCs for diverse areas	for tokens
Block time	15 s	15 s
Block confirmation	6 min	Instant (one block)
Throughput	10 tps	33 tps effectively (10 000 tps claimed)
Execution costs	Low to moderate (few $)	High (several thousand $)

Regarding the sustainability of the project, both projects depend on an active
ecosystem of developers and other contributors to succeed [19]. The community

of Ethereum is much larger, more international and exists for a longer time than Neo's. Both projects are well funded through ICOs [13,25].

Concerning dependence on companies or governments, Ethereum is fairly independent. The core developers eventually decide about the development. Neo has a similar dependency on its founders. The relationship between the Neo Foundation, OnChain and the Chinese government has never been clarified.

5   Conclusion

Our comparison showed, that Neo is a viable choice when (i) targeting the Chinese and related markets, (ii) energy consumption is a concern, (iii) dependence on the Chinese government is no concern, (iv) centralization is no concern, and (v) use cases are limited to the 'smart economy'. Its poor documentation and high execution costs make it virtually indispensable to cooperate with the Neo foundation.

Ethereum is a comparatively mature and decentralized general purpose solution without preference for any country or markets, albeit with a high energy footprint. The switch to Ethereum 2.0 may be a game changer.

Limitations. We took aspects of a platform as cryptocurrency into account only to the extent that they are relevant for the execution of SCs. Moreover, we did not aim at an in-depth performance test, but focused on functionality.

Future work. Regarding Neo, the claimed performance has not yet been confirmed due to the high costs of operation. Moreover, its relationship to the government of China should be reassessed. As both platforms are evolving, evaluating future versions seems worthwhile.

References

1. Alsunaidi, S.J., Alhaidari, F.A.: A survey of consensus algorithms for blockchain technology. In: International Conference Computer and Information Sciences (ICCIS). IEEE (2019)
2. Bareis, M.: Comparison of Ethereum and NEO as smart contract platforms. Master's thesis, TU Wien (2020)
3. Bartoletti, M., Pompianu, L.: An empirical analysis of smart contracts: platforms, applications, and design patterns. In: Brenner, M., et al. (eds.) Financial Cryptography and Data Security. LNCS, vol. 10323, pp. 494–509. Springer, Cham (2017)
4. Bocek, T., Stiller, B.: Smart contracts - blockchains in the wings. In: Linnhoff-Popien, C., Schneider, R., Zaddach, M. (eds.) Digital Marketplaces Unleashed, pp. 169–184. Springer, Heidelberg (2018)
5. Bosu, A., Iqbal, A., Shahriyar, R., Chakraborty, P.: Understanding the motivations, challenges and needs of blockchain software developers: a survey. Empirical Softw. Eng. **24**(4), 2636–2673 (2019)
6. Buterin, V.: Ethereum Proof-of-Stake FAQs. https://github.com/ethereum/wiki/wiki/Proof-of-Stake-FAQ. Accessed 29 Sept 2019

7. Coelho, I., Coelho, V., Lin, P., Zhang, E.: Community Yellow Paper: A Technical Specification for NEO Blockchain. https://neoresearch.io/assets/yellowpaper/yellow_paper.pdf. Accessed 07 Jan 2020
8. ConsenSys: Homepage. https://media.consensys.net. Accessed 05 Dec 2019
9. Cuen, L.: Neo relaunch. https://www.coindesk.com/neo-releases-detailed-financials-ahead-of-cryptocurrency-relaunch. Accessed 19 Aug 2019
10. Enterprise Ethereum Alliance. https://entethalliance.org. Accessed 11 Nov 2019
11. Ethereum Foundation. https://www.ethereum.org/. Accessed 15 Aug 2019
12. Ethereum Wiki. https://github.com/ethereum/wiki/wiki. Accessed 15 Aug 2019
13. EthHub. https://docs.ethhub.io. Accessed 8 Nov 2019
14. Fairley, P.: Ethereum plans to cut its absurd energy consumption by 99 percent. IEEE Spectr. **56**(1), 29–32 (2019). https://doi.org/10.1109/MSPEC.2019.8594790
15. FlatOutCrypto: ConsenSys: understanding one of the most important firms in crypto. https://hackernoon.com/consensys-understanding-one-of-the-most-important-firms-in-crypto-7e1d66533d4a. Accessed 28 Aug 2019
16. Harris, C.G.: The risks and challenges of implementing ethereum smart contracts. In: International Conference on Blockchain and Cryptocurrency (ICBC), pp. 104–107. IEEE (2019)
17. ICO Rating: Smart Contract Platforms Review. https://icorating.com/pdf/56/1//Bd4ljAOmjaCFAXmkCj9NAKZAIUj1Dwb9v75AAZe.pdf. Accessed 10 Dec 2018
18. Levenson, N.: NEO versus Ethereum. https://hackernoon.com/neo-versus-ethereum-why-neo-might-be-2018s-strongest-cryptocurrency-79956138bea3. Accessed 19 Aug 2019
19. Lyons, T., Courcelas, L., Timsit, K.: Scalability, interoperability and sustainability of blockchains. Technical report, European Union Blockchain Observatory & Forum (2019)
20. NEO Docs. https://docs.neo.org/docs/en-us/index.html. Accessed 02 Dec 2019
21. NEO release statement. https://neonewstoday.com/general/neo-release-statement-on-organisation-restructure/. Accessed 28 Aug 2019
22. NEP. https://github.com/neo-project/proposals. Accessed 04 Aug 2019
23. OnChain: Interview with Da Hongfei. https://www.youtube.com/watch?v=BcmoSp7bL7g. Accessed 28 Aug 2019
24. Parizi, R.M., Amritraj, A.D.: Smart contract programming languages on blockchains: an empirical evaluation of usability and security. In: Chen, S., Wang, H., Zhang, L.J. (eds.) International Conference on Blockchain, pp. 75–91. Springer, Cham (2018)
25. Rosic, A.: What is NEO Blockchain? https://blockgeeks.com/guides/neo-blockchain/. Accessed 18 Aug 2019
26. See, R.: The company behind Neo. https://hackernoon.com/neo-onchain-and-its-ultimate-plan-dna-4c33e9b6bfaa. Accessed 28 Aug 2019
27. Seth, S.: Why NEO Can Do What No Other Cryptocurrency Can Do. https://www.investopedia.com/tech/china-neo-cryptocurrency/. Accessed 15 Aug 2019
28. Ethereum: A Next-Generation Smart Contract and Decentralized Application Platform. https://github.com/ethereum/wiki/wiki/White-Paper. Accessed 15 Aug 2019

29. Wood, G.: Ethereum: a secure decentralised generalised transaction ledger. https://ethereum.github.io/yellowpaper/paper.pdf. Accessed 15 Aug 2019
30. Xu, X., Weber, I., Staples, M.: Architecture for Blockchain Applications. Springer, Heidelberg (2019)
31. Zheng, Z., Xie, S., Dai, H.N., Chen, W., Chen, X., Weng, J., Imran, M.: An overview on smart contracts: challenges, advances and platforms. Future Gener. Comput. Syst. **105**, 475–491 (2020)

Sandbox for Minimal Viable Governance of Blockchain Services and DAOs: CLAUDIA

Ismael Arribas[1], David Arroyo[2], and Denisa Reshef Kera[3]([✉])

[1] Kunfud (CCO at CLAUDIA), Parque Científico de la Universidad de Valladolid, Paseo de Belén 9A, 47011 Valladolid, Spain
kunfud@gmail.com
[2] Institute of Physical and Information Technologies (ITEFI), Spanish National Research Council (CSIC), Madrid, Spain
david.arroyo@csic.es
[3] BISITE, University of Salamanca, Edificio I+D+i - C/Espejo s/n, 37007 Salamanca, Spain
denisa.kera@usal.es

Abstract. To address the challenges of on-chain and off-chain governance of a blockchain project, CLAUDIA combines an on-chain services, such as Ethereum based DAO (Decentralized Autonomous Organization), timestamping solution (WUDDER), and decision making support (ARAGON), with off-chain "compliance desk". The off-chain "compliance desk" enables stakeholders to report security issues and breaches of the codes of conduct and discuss necessary actions. The robust solution makes CLAUDIA user friendly but also adaptable. It can be offered as an app to various publics that do not necessarily care about the underlying solutions, but it can also meet the needs of variety of institutions that operate under different jurisdiction and business models. We discuss the advantages of this hybrid solution that combines the current modes of blockchain application testing in testnets and sandboxes to tackle the challenges of blockchain governance.

Keywords: DAO · Minimal viable governance · Sandbox · Blockchain governance

1 Introduction

There is a tension between the on-chain and off-chain governance of blockchain services, between the ideal of a trustless and anonymous network fully managed by an algorithm (consensus mechanism, hashgraph) and a need for maintenance, management, and upgrade of the actual software and infrastructure [6, 7, 14, 15]. While the on-chain governance is performed by nodes, peers, and what Rachel O'Dwyer describes as "non-specific CPU - producing entities" [12] guided by UNIX epoch clocks cycles, the actual maintenance of the platform is performed by humans, communities and various stakeholders with agendas and interests.

This tension is often simplified as a matter of choice of the right consensus algorithm rather than a problem between governance based on code versus governance of

J. Prieto et al. (Eds.): BLOCKCHAIN 2020, AISC 1238, pp. 24–30, 2020.
https://doi.org/10.1007/978-3-030-52535-4_3

code [2]. In most cases, security flaws or maintenance of the blockchain systems lead to rather arbitrary exercises of power by part of the developers or core stakeholders creating new hard-forks or upgrading the blockchain with cascading effects that erode the trust [1, 4, 9]. The software ideals of automation, neutrality, transparency or agnosticism performed by the "code" and algorithms behind the system clash with the actual power relations between the stakeholders of the platform that design, maintain and upgrade these algorithms [16].

To address these challenges of connecting the on-chain and off-chain governance of blockchain projects, the Ethereum based DAO (Decentralized Autonomous Organi-zation) CLAUDIA[1] combined its autonomous governance over blockchain solution for timestamping (WUDDER)[2] and decision making (ARAGON)[3] with an off-chain "compliance desk" that enables stakeholders to report security issues and breaches of the codes of conduct. This combination of code and off-chain "desk" enables the community to discuss in parallel the necessary on-chain and off-chain actions to pre-vent further problems. The robust solution makes CLAUDIA not only user friendly (it can be offered as an app to various publics that do not care about underlying solutions), but also well adapted to existing institutions that operate under various jurisdiction (it supports the "Limited Liability Corporation" layer) and business models.

In this sense, CLAUDIA follows and extends the hybrid model of the regulatory sandboxes for testing new applications on a limited number of users over limited time. The purpose of a sandbox is to identify the changes needed in terms of both code and regulations which the compliance desk on the top of the smart contracts performs in parallel and works in synchronized fashion on both the code and regulations. The CLAUDIA compliance desk extends the quarantined sandbox model by com-bining the ideals of a "working code and rough consensus" [5, 11] as "humming" [13] and the agile inspired "minimal viable governance" [17, 18] model.

Governance then becomes a matter of actionable pragmatic choices, on which the community can cooperate without excluding stakeholders that do not program or do not have access to the core developers' team. This hybrid solution both extends the sandbox and testnets models but also offers an alternative to the automated solutions that are vulnerable to cascading errors and security issues.

2 From Testnets to Sandboxes for Humming

The hybrid, on-chain and off-chain CLAUDIA governance platform enables the vari-ous stakeholders to interact over code, but also off-chain contracts and codes of conduct and identify changes needed to prevent unintended consequences, security flaws, and other risks. The solutions combines the two testing models and environments currently in use in blockchain development: testnets [8, 10] and sandboxes [3].

---

[1] https://github.com/kunfud/CLAUDIA.

[2] https://github.com/wuddertech/.

[3] https://aragon.org/.

On one side, the various testnet, such as Ethereum's Rinkeby[4] or Ropsten[5], support the developers and miners to test the changes and applications on the blockchain or decentralized ledger and identify flaws or incompatibilities. On the other side, regulatory sandboxes commonly used for FinTech and RegTech solutions, work with "off-chain" regulations and include the users and stakeholders in the testing that improves both the technology and, to various degrees, the regulations.

CLAUDIA represents a hybrid type of sandbox and testnet that engages all stakeholders that use, design, but also care about the codes of conduct related to the project's infrastructure. It includes a reporting mechanism which is off-chain and serves to improve the actual code and technology based on the codes of conduct set up together with the platform. In the present, there was one incident reported that lead to a proposal to an actual change of the code that made the platform safer and more trustworthy.

What differentiates such hybrid sandbox for blockchain governance from the usual testnets or regulatory sandboxes is the impossibility to reduce it to only one tool, goal or agenda that serves only one type of stakeholders. It engages diverse stakeholders that create and experiment with finding a common ground for action and exchange (or "humming") by reporting early on the issues and acting upon them.

Governance and Interoperability
CLAUDIA is composed with the following layers:

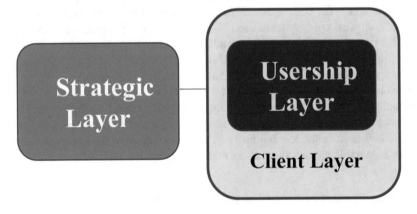

Client Layer: includes the Usership Layer, full interoperability, collective intelligence data belongs to the Client and it is provoked for client's purposes. Who is a user in CLAUDIA? Anyone who can have access to the participation by Client's choice. Who is stakeholder of CLAUDIA? Anyone who proof DAO participation.

Strategic Layer: CLAUDIA is allocated on wudder, a layer 2.0 on Ethereum main-net with functionalities for the client, it is case by case dealt, with three minimal automated

---

[4] https://www.rinkeby.io/.

[5] https://faucet.ropsten.be/.

resources thru a GraphQL API to interact with HTTP API and a variety of SDKs: time-stamping of the openness, time-stamping of closing and certification of results, with global average results, using master token keys hash function (MTK-512, which is a combination of BLAKE2b with 512-bit output and SHA3-512). Certification can be enforced within various model and trust anchoring simultaneously in various DLTs and Blockchains, these circumstances depend on the scenario of integration or application. This layer is a basic an extensible vehicle for scalability purposes and it is under a Policy for Unified Incident Response Management Protocol.

Usership Layer: It's a very simple web-app which did not request any data or PII (Personal Identifiable Information) from the user. Thru a simple lector incorporated on the web-app the user accesses to the survey or consultancy. Privacy by default and design is a key ingredient in compliance with GDPR.

Unified Incident Response Management Protocol
As an observer CLAUDIA has identified a variety of use cases described below in the figure, however a compliance reporting mechanism for off-chain incidents is in nature of CLAUDIA as a duality, one multi-DAO based on compartmental and an MvP to survey opinions of different material. It is a governance tool for interoperability and it has been created and adheres the scheme for optimal response with an Incident management situation:

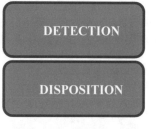

DETECTION

User confirmation or communication.
Client confirmation. Internal confirmation.

DISPOSITION

Issue Documentation and Library and Wiki of CLAUDIA. Compliance Desk protocol implementation

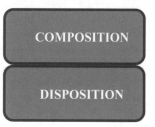

COMPOSITION

Assistance for integration. Chat-bot and web-bot capabilities. Compliance reporting.

DISPOSITION

Publication and CLAUDIA Improvement Proposal Issuance.

CLAUDIA categorizes three main areas of incident monitoring:

1. Irregular processing.
2. Fraudulent processing.
3. Irresponsible processing.

3   CLAUDIA Connecting DAO Automation with Sandbox Iteration

CLAUDIA is a compartmental community that uses blockchain technology to build applications for consultation processes and surveys in events and meetings. It supports anonymity and fairness of opinions (prevents cheating) while preserving the principles of privacy by design through anonymization and pseudo-anonymization. The users remain sole data owners of their opinions and the consultations' results.

CLAUDIA enables the configuration of the surveys with a maximum of four questions, which is a reasonable trade-off between optimal and objective attention and users ongoing activity. It allows the users to decide when and how they will share the outputs and what are the rewards: "Collective Intelligence is our service and CLAUDIA is our MvP (minimum viable product) for you to rise the voice anonymously and being rewarded for your fairness in your opinion." (Internal documents available also on Github of the project).

This community is built as a DAO with a compliance desk and comprises "departments," like CTO (Chief Technical Office), CEO (Chief Executive Office) or CCO (Chief Compliance Office). The governance model is startup based as a MvP in a form of a web-app for supporting the business model sustainability. On one side, CLAUDIA is a decentralized community and on the other, an application to survey different areas of organizations, governmental administrative office, NGOs, communities, businesses, events. It can also enable a variety of compliance testing via the web-app with which traces in an anonymous definition the public opinion and generate a reporting of compliance off-chain based. In short, it understands off-chain and on-chain governance as balancing automation with reflection, feedback, and iteration processeses.

4   Development of CLAUDIA

CLAUDIA was incubated under a university program named TIMMIS in the University of Valladolid and started as a seed-blockchain project in competition at Startup Olé 2019 in Salamanca, Kingdom of Spain, whereby was rewarded in such category by OurCrowd and AWS. In 2019 CLAUDIA tested the hybrid governance model of a DAO by integrating various on-chain and off-chain governance models.

In the first stage, the Governance Guidelines (codes of conduct) focused on the incident responsiveness and the assertivity of their minimal viable governance as a multiDAO. The facets of the system were designed for general purpose to support specific purposes such as payments or other specific industry needs. Like most DAO governance efforts, in this phase the project just replicated the off-chain governance with on chain tools by using cryptography and ciphering their data within different DAOs. What CLAUDIA added to this, was the Compliance Desk, which is functioning as an automated responsiveness mechanism, because it was committed to compliance reporting since its inception to serve the need for transparency in the events and activities that are participating.

Second ingredient is the extensibility, CLAUDIA tested a first survey within a private event at SEMINCI 64 which is an international film festival famous within the Indy and Author section with around 600 participants. The second reiteration still as PoC was for INATBA (the International Association of Trusted Blockchain Applications) event for more than 1400 attendees, just an extensibility testing, and the third was for a Cultural festival named Cultural over 14000 attendees, for final modeling.

The project proofed the ability of anonymization and pseudonymization with privacy by default. Basics are on MTK-512 which is a hybrid algorithm between Blake2b and SHA-3 for 512 bits on the client's side a simple ERP to parameterize CLAUDIA on their business consults and/or events, with two layers: filtering layer and communication layer.

Through a variety of SDKs and APIs generates three robust digital evidences on WUDDER which is a sidechain on Ethereum as a trust ledger 2.0 for CLAUDIA in each appearance to the vox populi: opening of the consult, closing of the consult and results (average results). Note that in one of their PoCs were included a four element as Smart Contract to reward one participant of the survey. The third part is adaptability, CLAUDIA is testing the integration of voice and tools of Artificial Intelligence for future purposes. Example of that is COP25 CCO's CLAUDIA was invited to join as a volunteer for a scientific task force which is focused on green reporting, the initiative is named greenfilling and CLAUDIA will be parameterized to calculate various Carbon´s Footprint and serving the green reporting with mechanism for taxonomy and governance.

5   Future of CLAUDIA

The project is currently trying to scale in Switzerland, Italy, Cyprus, Germany, Austria and France and its main goal is the SDG 10 "Reducing inequalities". It is a member of a global alliance named Lumiversity® an ICT (Information and Communications Technology) entrepreneurship for social impact initiative by ITU (The International Telecommunication Union). It is also trying to integrate the Global Navigation Satellite System (Galileo Constellation) to tackle the challenge on tracking things and not only organizations and using traceability and artificial intelligence in decision making to support responsiveness mechanisms for off-chain compliance. The combination of on-chain and off-chain procedures and tools helps this integration of various technologies into the blockchain ecosystem to support decentralization and (pseudo)anonymization while keeping compatibility with the applicable regulation and accountability against competent and relevant off chain institutions.

References

1. Anceaume, E., Lajoie-Mazenc, T., Ludinard, R., Sericola. B.: Safety analysis of Bitcoin improvement proposals. In: Proceedings of 2016 IEEE 15th International Symposium on Network Computing and Applications, NCA 2016 (2016)

2. Andrés, P., Arroyo, D., Correia, R.: Governance institute-law, and undefined 2019: regulatory and market challenges of initial coin offerings. ECGI Working Paper Series in Law (2019)
3. Bromberg, L., Godwin, A., Ramsay, I.: Fintech sandboxes: achieving a balance between regulation and innovation. J. Bank. Financ. Law Prac. **28**(4), 314–336 (2017)
4. Decker, C., Wattenhofer, R.: Information propagation in the Bitcoin network. In: Proceedings of 13th IEEE International Conference on Peer-to-Peer Computing, IEEE P2P 2013 (2013)
5. DeNardis, L.: Protocol Politics: The Globalization of Internet Governance. MIT Press, Cambridge (2009)
6. De Filippi, P., Hassan, S.: Blockchain technology as a regulatory technology: from code is law to law is code. First Monday **21**, 12 (2016)
7. De Filippi, P., Loveluck, B.: The invisible politics of Bitcoin: governance crisis of a decentralised infrastructure (2016)
8. Hertig, A.: Immature code or good test? Bitcoin scaling proposal Segwit2x's testnet forks. Coindesk (2017). https://www.coindesk.com/bitcoin-segwit2x-testnet-fork-scaling-proposal. Accessed 21 Jan 2020
9. Karame, G.O.: On the security and scalability of Bitcoin's blockchain. In: Proceedings of the ACM Conference on Computer and Communications Security (2016)
10. Kaur, J.: 10 Blockchain simulators and testnets for all your testing needs. Hackernoon (2020). https://hackernoon.com/blockchain-simulators-ui2030z0. Accessed 21 Jan 2020
11. Denardis, L.E., Abbate, J., Allen, B., Downey, G., Hauger, J.S., Hirsh, R.: IPv6: Politics of The Next Generation Internet (2006)
12. O'Dwyer, R.: Code ! = Law: Explorations of the Blockchain as a Mode of Algorithmic Governance (2018)
13. Resnick, P.: On consensus and humming in the IETF, pp. 1–19 (2014) https://tools.ietf.org/id/draft-resnick-on-consensus-01.html. Accessed 6 Feb 2020
14. Rikken, O., Janssen, M., Kwee, Z.: Governance challenges of blockchain and decentralized autonomous organizations. Inf. Polity **24**(4), 397–417 (2019)
15. Trump, B.D., Wells, E., Trump, J., Linkov, I.: Cryptocurrency: governance for what was meant to be ungovernable. Environ. Syst. Decis. **38**(3), 426–430 (2018)
16. Lehdonvirta, V.: The blockchain paradox: why distributed ledger technologies may do little to transform the economy. Oxford Internet Institute blog (2016)
17. Steps towards minimal viable governance | APM. https://www.apm.org.uk/blog/3-steps-towards-minimal-viable-governance/. Accessed 6 Feb 2020
18. Introducing Minimal Viable Governance | LeapDAO. https://leapdao.org/blog/Minimal-Viable-Governance/. Accessed 6 Feb 2020

A Study on Recent Trends of Consensus Algorithms for Private Blockchain Network

Prasad B. Honnavalli$^{(\boxtimes)}$, Ajaykumar S. Cholin, Athul Pai,
Achuta D. Anekal, and Aditya D. Anekal

Center for Information Security, Forensics and Cyber Resilience,
PES University, 100 Feet Ring Road, BSK III Stage,
Bengaluru, Karnataka, India
prasadhb@pes.edu, cholinajay@gmail.com,
athulpai05@gmail.com, achutaanekal@gmail.com,
adityaanekal@gmail.com

Abstract. Blockchain is expected to play a seminal role in the future of finance and cybersecurity. Blockchain has the traits of decentralization, stability, safety, and non-modifiability. The consensus algorithm performs an important role in retaining the order of transaction and performance of blockchain networks. Using the right algorithm might also deliver a great boom to the performance of the blockchain applications. In this paper, we reviewed the ideas and characteristics of some of the different consensus algorithms and analyzed the performance and application eventualities of various consensus mechanisms with an example of an electronic voting system. The electronic voting system needs the ability to address the following constraints: Performance, Fault Tolerance, Byzantine Fault Tolerant to make it a scalable application. We have also studied different algorithms, in particular, the ones in distributed systems and came up with a novel way to solve the problem mentioned above.

Keywords: Blockchain · Fault tolerance · Consensus · PBFT · CFT · IBFT · POA

1 Introduction

The concept of blockchain was first introduced to the world by Haber and Stornetta as a time-stamped digital document [1]. In 2008, Satoshi Nakamoto introduced Bitcoin, a peer-to-peer electronic cash system [2]. The underlying technology for Bitcoin was known as a blockchain. This happened to be a radically different method of data storage and accessibility, then the world had ever witnessed. A blockchain is, in simplest of terms, a time-stamped series of an immutable record of data that is distributed among not one entity but, a cluster of computing systems. Each block of data is secured and bound to each other using cryptographic functions forming a chain of blocks. This distributed ledger is managed by a peer-to-peer network. The popularity of blockchain has been bolstered in recent times with the advent of many use cases other than cryptocurrency. Some of them are smart contracts, record and identity management, supply chain management, Dapps, banking, healthcare, to name a few. For a blockchain network to

J. Prieto et al. (Eds.): BLOCKCHAIN 2020, AISC 1238, pp. 31–41, 2020.
https://doi.org/10.1007/978-3-030-52535-4_4

work, all nodes/entities involved agree over the working of the network. This agreement is achieved through a consensus algorithm that governs the working of the blockchain network. This consensus is essential for verifying transactions and adding blocks to the ledger by the participating nodes. Some of the consensus algorithms in use are Proof of Work (Bitcoin), Proof of Stake, and voting based ones like PBFT in case of Hyperledger Fabric which is a permissioned blockchain network. Hyperledger is an open-source project with various blockchain technologies under its wings [3]. This research paper aims to describe different consensus algorithms, its particular use case and implementation, and our inferences from this study. An example of a real-world use case like voting using these consensus algorithms is explained later in Sect. 5.

2 Background

2.1 Permissionless Blockchain

Much like how the internet was created in the 1980s and 90s, permissionless blockchain [4] will allow anyone to access data, create data, write smart contracts and run their code to take part in the network. This type of blockchain provides 100% transparency and a relatively high level of anonymity. Because it's an open community with public ownership, permissionless blockchain permits anyone to create blocks and leverage complex algorithms which brings about performance and scalability hurdles as it requires tremendous amounts of computing power. However, permissionless blockchain is the best bet to provide a completely decentralized, distributed and secure database. Examples such as Bitcoin, Ethereum and other crypto-economic businesses, which run a peer to peer, trustless system enjoy the framework of permissionless blockchain.

2.2 Permissioned Blockchain

A permissioned blockchain [4] defines a closed ecosystem where all the participants are defined. Only pre-approved entities can run the nodes. It provides a multilayer information sharing function that is necessary for a private community. The private community will decide the levels of decentralization and transparency it needs, which gives the luxury of configurational flexibility. There is no mining to validate transactions or execute smart contracts. It creates a trust-based system with no anonymity as all user's identity is pre-approved. Unlike the case of Bitcoin and Ethereum, there is no token for the running nodes. All these benefits result in high performance with low computing applications, allowing scalable networks for example Hyperledger framework.

2.3 Consensus Algorithm

The fundamentals for blockchain networks to function, is to have some sort of consensus between the participants [5]. A consensus also known as agreements, can be by modifying or appending a new block or verifying the transaction between two participants. A consensus is done in a peer to peer network to agree on the state of the network. There are two types of consensus algorithms. One is proof-based consensus algorithms such as Proof of Work (PoW), Proof of Stake (PoS), Proof of Authority (PoA) [6].

The second type of consensus algorithm is called- voting-based consensus algorithm. To execute this algorithm, the nodes inside the network should be known and adjustable, so that they can exchange the message easily. The main difference when compared to proof-based consensus algorithms, is that nodes are often free to join and withdraw from the network. Executing voting-based consensus algorithms is similar to traditional methods of tolerating faults used in the distributed system. Therefore, voting based consensus should be designed to resist some bad cases. Some nodes will subvert or crash. Based on these bad situations, the voting-based consensus algorithms can be classified into two main kinds; (i) Byzantine fault tolerance based consensus- a kind of consensus that could prevent the cases of crashing nodes and subverted nodes; (ii) Crash fault tolerance based consensus- a kind of consensus that could only prevent the crashing nodes. The voting-based consensus algorithms described in this research are: Practical byzantine fault tolerance. Notable projects to use BFT are Hyperledger and Zilliqa. Another is Istanbul byzantine fault tolerance and an example this is AMIS and the last voting-based consensus algorithm explained is Crash Fault Tolerance.

3 Explanation of the Different Consensus Algorithm

3.1 Practical Byzantine Fault Tolerance

Byzantine Fault is a condition in distributed computing systems where, systems fail and there is inconsistent information on whether it has failed. This is derived from- Byzantine Generals Problem. It describes this situation and a strategy to be followed by actors, to avoid catastrophic system failure, when there are malicious actors involved [14].

Practical Byzantine Fault Tolerance ensures the liveness and safety of a network of computing systems even when some portion of the network may be malicious or faulty [12, 15]. In the case of a blockchain network, each node is a computing system. As long as a minimum percentage of nodes in the PBFT blockchain network are connected, working properly, and behaving honestly, the network will always make progress and will not allow any of the nodes to manipulate the network [13]. PBFT is a permissioned consensus algorithm in blockchain, which means the network is private and only upon explicit consent from the other nodes can a node enter the network [14].

A PBFT network consists of several nodes which are ordered from 0 to $n - 1$. There is a maximum number of 'faulty' nodes that the network can tolerate. As long as this maximum number of 'faulty' nodes (represented by f) is not exceeded, the network will function properly. In PBFT it should not be more than 1/3rd of the whole network [12, 14]. Maximum number of Faulty Nodes,

$$f = n - 1/3 \tag{1}$$

For Non-byzantine failures:

Given n nodes, with f faulty nodes, the Quorum (minimum number of honest nodes) size Q required to guarantee liveness and safety of the network is given by:

$$\text{For, liveness } Q \leq n - f \tag{2}$$

$$\text{For, safety } Q > n/2 \tag{3}$$

Combining the above Eqs. (2) and (3) we get,

$$n > 2f$$

$$\text{If, } n_{min} = 2f + 1, \text{ then, } 2Q > f + 1$$

Therefore, for Non-Byzantine failures,

$$Q_{min} = f + 1 \tag{4}$$

For Byzantine failures:

Given n nodes in a network, with f fault nodes which might experience Byzantine failure, the Quorum size Q required to guarantee liveness and safety of the network is given by:

$$\text{For, liveness } Q \leq n - F \tag{5}$$

$$\text{For, safety } 2Q - n > F \tag{6}$$

Combining the above Eqs. (5) and (6) we get,

$$n > 3f$$

$$\text{If, } n_{min} = 3f - 1, \text{ then, } 2Q > 4f + 1$$

Therefore, for Byzantine failures

$$Q_{min} = 3f + 1. \tag{7}$$

PBFT has three key parts as follows: Normal-Case Operation, Garbage Collection and View Change [12]. Normal-Case Operation enables requests from nodes to be executed in a certain order. The Garbage Collection regularly generates system log files for recovering, in case of a fault. The View Change will be executed when the primary node has added a block to the network successfully and then a new primary will be elected. A primary node for each view is elected in a round robin fashion.

The primary (p) for each view is determined based on the view number (v). The formula for determining the primary for any view on a given network is [14]:

$$p = v \bmod n \tag{8}$$

For example, on a 5-node network at view 8, the formula $p = 8 \bmod 5$ means that node 3 will be the primary.

In addition to moving through a series of views, it also moves through a series of sequence numbers. In the context of a blockchain network, a sequence number is equivalent to a block number. To commit to a block and make progress in the network, the nodes in a PBFT network go through these three phases [16, 17]:

Pre-preparing Phase. The primary for the current view will create a block and publish into the network. Each node will receive this block and perform some initial verification to make sure that the block is valid. Next, it broadcasts a pre-prepare message to all of the nodes. Pre-prepare messages contain four key pieces of information: the ID of the block just published, the block's number, the primary's view number, and the primary's ID [17]. Validating nodes wait for 2f + 1 valid pre-prepare messages and enter the next stage.

Preparing Phase. Here, the node will broadcast a prepared message to the rest of the network (and itself). Prepare messages are very similar to pre-prepared messages. To move to the next phase, the node must wait until it has received 2f + 1 prepare messages that have the same block ID, block number, and view number, and are from different nodes. By waiting for 2f + 1 matching prepare messages, the node can be sure that all properly functioning nodes (non-faulty and non-malicious) agree at this phase [17].

Committing Phase. Here, a node entering this phase will broadcast a commit message which is identical to the previous message type. In preparing phase, a node cannot complete the committing phase until it has received 2f + 1 matching commit messages from different nodes. Again, this guarantees that all non-faulty nodes in the network have agreed to commit this block, that is, the node can safely commit the block knowing that it will not be reverted. With the required 2f + 1 commit messages accepted and, in its log, the node can safely commit the block [17]. Once this done the whole process will start again.

3.2   Istanbul Byzantine Fault Tolerance

IBFT is one of the consensus algorithms that ensures that the transactions have one sole truth and has the ability to support the need for enterprise-scale and speed. IBFT is one that does not have the scale to reach public blockchain platforms but in the realm of a private blockchain, it is very popular.

The Proof of Work algorithm (PoW) calculation is broadly expensive, in both equipment and power. This expense is purposeful, to prevent people from effectively assuming control over the system. Proof of work, in the same way, is pretty expensive in terms of the resources consumed.

Considering the above constraints IBFT [20] comes out as a possible solution to the problem of resource consumption on a private network. IBFT ensures working conditions given the fact that less than one-third of the validating nodes are invalid.

The working of IBFT is as shown below (Fig. 1).

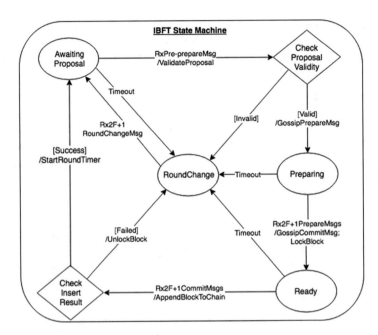

Fig. 1. IBFT state machine

IBFT gives substantial benefits when used on a private network where the validator pool is trusted and held responsible for the transaction processing.

3.3   Crash Fault Tolerance

Crash Fault Tolerant [CFT] consensus algorithm [18] is a voting-based consensus algorithm [4]. Executing voting-based consensus algorithms is like traditional methods for tolerating faults used inside the distributed systems [19]. Therefore, vote casting-based consensus must be designed to encounter some bad instances such as crashed nodes or nodes that crash and subvert.

In case of crashing, nodes will look forward to the messages from other nodes. However, there are some nodes that don't run, in which case the normal nodes do not receive enough evidence to make the decision. Therefore, to prevent any instance of crashing with f nodes, there must be at least f + 1 nodes running normally. Typically voting-based consensus is classified into two main kinds: (a) BFT based consensus, (b) CFT based consensus.

$$t = [(N/2) + 1] \tag{9}$$

$$t = [(2N/3) + 1] \tag{10}$$

In any consensus algorithm generally, let us assume among N nodes, there ought to be at least t nodes (t < N) operating normally. While in crash fault tolerance-primarily based consensus, 't' is generally set equal to [N/2 + 1] as in Eq. (9). In byzantine fault tolerance-based consensus 't' is commonly assigned equal to [2N/3 + 1] as in Eq. (10). Hence in this context the CFT consensus algorithm is suitable in real time applications that needs to be highly robust to fault tolerance without taking subverted nodes into consideration.

3.4 Proof of Authority

For Bitcoin, we use a reliable and secure version of a consensus algorithm, which is Proof of Work (PoW) [2]. In PoW consensus algorithm, a node has authorized consensus to share its blocks only when enormous computing power has been exerted from its side. Which means, a central authority monitors the transactions between the nodes and has to sign it (authorize it) to allow the transaction. This results in issues like high computing power, higher spending, censorship problem (A party does not act appropriately if he dislikes the other party) and vulnerability to a 51% attack.

A partial solution to this is Political Mitigation, which is politically decentralizing the authority. This means creating individual authorities to every single independent node. Now they can all maybe represent different interests of different parties. All the nodes have to get together to sign off on any transactions, this is called MultiSig [7]. Although it reduces the problems faced above, there is no open access and no in-protocol penalties. In Bitcoin, to reach consensus, we have to get verified and approved by a majority of the active nodes in the network before a transaction is made. Proof of Stake (PoS) [11] although providing better performance by appointing stake to each node, does not solve the problem of scalability like how Proof of authority does.

An approach to overcome the drawbacks of PoW and PoS is something known as Proof of Authority. Proof of Authority is a consensus mechanism that is based on the process of using the identities of the network's users to establish the best validator for a transaction [8]. This creates a reputation-based consensus system that helps us build an efficient and practical network and allows up to 50% malicious nodes.

In the PoA consensus algorithm, instead of staking the block validators coins during a transaction, its reputation is at risk. This means that PoA values identities of the nodes, making its blockchain secure and allowing trustworthy entities (arbitrarily selected) to validate the nodes. Here, algorithms rely on a set of N trusted nodes called the authorities which are randomly selected by a trustworthy entity. A unique ID is allotted to each authority by which it is identified. There should be at least N/2 + 1. Once a client orders the necessary transactions, a consensus is run by the authorities.

Conditions that must be fulfilled for a validator to be established are [9]: (i) Identity must be formally verified on-chain, with a possibility to cross-check the information in a publicly available domain; (ii) Eligibility must be difficult to obtain, to make the right to validate the blocks earned and valued (potential validators are required to obtain public notary license). (iii) There must be complete uniformity in the checks and procedures for establishing an authority.

The consensus in PoA algorithms works on a mining rotation principle [4] which is a widely used approach that helps in fairly distributing the responsibility of block creations among the selected authorities [10].

The essence behind the reputation mechanism is the certainty behind a validator's identity. Ensuring that all validators go through the same procedure guarantees the system's integrity and reliability. Proof of Authority can be applied to a variety of applications that demand high-value options for logistical reasons to give us a more scalable model. Although PoA doesn't promise a completely decentralized network it gives us a fusion between decentralization and the efficiency that centralized networks enjoy.

4   Inferences

(See Fig. 2 and Table 1).

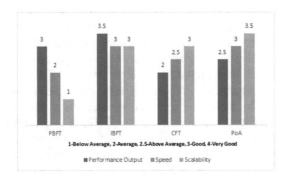

Fig. 2. From the above graph we can infer visually, how the four consensus algorithms compare against each other from our study.

Table 1. Shows us how the different performance criteria measure up against the four different consensus algorithms and what scenario it works best in.

Consensus algorithm	Performance type			Suitable scenarios
	Handling faulty nodes	Handling node failures	Efficiency/performance	
PBFT	GOOD upto 33%	GOOD	AVERAGE	Malicious nodes/components
PoA	GOOD upto 50%	ABOVE AVERAGE	GOOD	Where security is critical
IBFT	GOOD upto 33%	GOOD	GOOD	When performance is critical
CFT	ABOVE AVERAGE	GOOD	ABOVE AVERAGE	Hardware/software failures

5   Real World Use Case

Taking the problem of solving electronic voting system by using a blockchain system, we shall see how different consensus algorithms will play into effect. Electronic voting is not possible in a traditional client-server due to the enormous scale, introducing a single point of failure which is very hard to recover from when a large scale needs to be addressed. Electronic voting system needs the ability to address these constraints:-

1. *Speed:* Speed is needed as the application needs to have a fast and reliable way to provide a user to have a fast way to vote for the governing body of their choice.
2. *Fault tolerance:* The application needs to be fault-tolerant and should not have a single point of failure that would crash the system. The application needs to recover even after a crash seamlessly.
3. *Byzantine Fault Tolerant:* The application needs to work even when there is a presence of an adversarial body in the system. The application needs to perform seamlessly even when a malicious node is present in the system.
4. *Large Scale:* The application needs to scale linearly as the number of concurrent users using the system increases.

Initial Setup: The electronic voting system was set up using Azure blockchain as shown in Fig. 3. Docker images were used to simulate the real-world voters. The docker images were running in different AWS lambda servers.

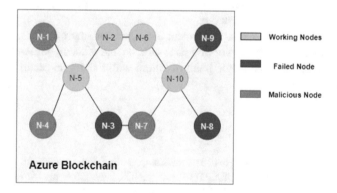

Fig. 3. Simulated Azure blockchain setup

Initial Architecture: The large blockchain network was configured to have 1000 random nodes and 10000 random voters simulated to use the system simultaneously. Different consensus algorithms were tested on this setup. The following scenarios are an extension of the initial architecture. Below are the results obtained using the above-mentioned setup. Figure 3 represents a sample 10 node network.

Scenario I: One-third of the nodes in the blockchain network was configured to crash every ten minutes. IBFT > PoA > CFT > PBFT.

Scenario II: The number of client nodes (i.e. the voters accessing the blockchain network) were linearly increased 100 folds. IBFT $\sim$ PoA > CFT > PBFT.

Scenario III: One-third of the nodes in the blockchain was tuned to be malicious over time and the following results were obtained. PBFT > PoA > IBFT $\sim$ CFT.

The ordering for the above scenarios is based on the summary of the studied consensus algorithms as shown in Table 1. IBFT turns out to be the most efficient in cases where nodes become faulty and crashes because in the Validation Group Proposal stage in an IBFT state machine, each node tracks the other nodes in real time making it aware of regrouping a network in case there is a crash of up to 33% of the nodes. Although PoA provides a robust authority system, its limitation on the number of block validators helps it to be as scalable as IBFT. In any transaction, PBFT handles malicious nodes the best as it uses a distributed network where at least 2f + 1 nodes should be approved.

6   Conclusions and Future Work

Understanding different consensus algorithms which are used in a distributed network is crucial for improving the performance of a blockchain network. Our paper focuses on clearly defining four major consensus algorithms used in a private blockchain network. Unlike traditional consensus algorithms, these four algorithms are designed for more optimal and superior performance. Using electronic voting as a simulated use case, we were able to interpret how the use of these studied consensus algorithms would play out in different scenarios.

With this study, we hope to implement a network of smart devices, all connected through a blockchain network working on a suitable and efficient consensus algorithm. This ensure that there is no single point of failure with maximum security and privacy in a decentralized manner.

References

1. Haber, S., Stornetta, W.S.: How to time-stamp a digital document. In: Menezes, A.J., Vanstone, Scott A. (eds.) CRYPTO 1990. LNCS, vol. 537, pp. 437–455. Springer, Heidelberg (1991). https://doi.org/10.1007/3-540-38424-3_32
2. Nakamoto, S.: Bitcoin: a peer-to-peer electronic cash system (2008)
3. Dhillon, V., Metcalf, D., Hooper, M.: The hyperledger project. In: Blockchain Enabled Applications, pp. 139–149. Apress, Berkeley (2017)
4. Nguyen, G.T., Kim, K.: A survey about consensus algorithms used in blockchain. J. Inf. Process. Syst. **14**(1), 101–128 (2018)
5. BitFury Group in Collaboration with Jeff Garzik Public vs Private Blockchains. https://bitfury.com/content/downloads/public-vs-private-pt1-1df
6. Basic Primer: Blockchain consensus protocol. https://blockgeeks.com/guides/blockchain-consensus/
7. Multisignature. https://en.bitcoin.it/wiki/Multisignature

8. POA Network Whitepaper. https://github.com/poanetwork/wiki/wiki/POA-Network-White paper
9. Proof of Authority. https://tokens-economy.gitbook.io/consensus/chain-based-hybrid-models/proof-of-authority-poa
10. De Angelis, S., Aniello, L., Baldoni, R., Lombardi, F., Margheri, A. Sassone, V.: PBFT vs proof-of-authority: applying the cap theorem to permissioned blockchain (2018)
11. Dziembowski, S., Faust, S., Kolmogorov, V., Pietrzak, K.: Proofs of space. In: Gennaro, R., Robshaw, M. (eds.) CRYPTO 2015. LNCS, vol. 9216, pp. 585–605. Springer, Heidelberg (2015). https://doi.org/10.1007/978-3-662-48000-7_29
12. Castro, M., Liskov, B.: Practical Byzantine fault tolerance and proactive recovery. ACM Trans. Comput. Syst. (TOCS) 20(4), 398–461 (2002)
13. Sukhwani, H., Martínez, J.M., Chang, X., Trivedi, K.S., Rindos, A.: Performance modeling of PBFT consensus process for permissioned blockchain network (hyperledger fabric). In: 2017 IEEE 36th Symposium on Reliable Distributed Systems (SRDS), Hong Kong, pp. 253–255 (2017)
14. Zheng, Z., Xie, S., Dai, H., Chen, X., Wang, H.: An overview of blockchain technology: architecture, consensus, and future trends. In: 2017 IEEE International Congress on Big Data (BigData Congress), Honolulu, HI, pp. 557–564 (2017)
15. Wang, X., WeiLi, J., Chai, J.: The research on the incentive method of consortium blockchain based on practical Byzantine fault tolerant. In: 2018 11th International Symposium on Computational Intelligence and Design (ISCID), Hangzhou, China, pp. 154–156 (2018)
16. Zhang, L., Li, Q.: Research on consensus efficiency based on practical Byzantine fault tolerance. In: 2018 10th International Conference on Modelling, Identification and Control (ICMIC), Guiyang, pp. 1–6 (2018)
17. Olson, K., Bowman, M., Mitchell, J., Amundson, S., Middleton, D., Montgomery, C.: Sawtooth: an introduction. The Linux Found (2018)
18. Lamport, L.: Paxos made simple. ACM Sigact News 32(4), 51–58 (2001)
19. Heimerdinger, W.L., Weinstock, C.B.: A conceptual framework for system fault tolerance. Defense Technical Information Center, Technical report CMU/SEI-92-TR-033 (1992)
20. Saltini, R.: Correctness analysis of IBFT. arXiv preprint arXiv:1901.07160 (2019)

A Fair and Anonymous Payment System for the Onion Relays

Debasish Ray Chawdhuri$^{(\boxtimes)}$

Talentica Software (India) Pvt. Ltd., Pune, India
debasish.chawdhuri@talentica.com

Abstract. Tor, aka 'The Onion Router', is a protocol that allows anonymous browsing of the web in the sense that the network address of the client is not known to a single relay or even relays provided by the same provider. The relay that knows the target does not know the source and vice versa. However, the current Tor relays are voluntary and unpaid, doing a social service to protect the identities of people who want to stay anonymous while browsing the web. We propose a modification of the Tor protocol so that the client can anonymously pay the Tor relays for their services through a smart-contract enabled cryptocurrency. Our design ensures that the Tor relays cannot be paid unless the client gets the data and that the client cannot get the data without paying the Tor relays.

Keywords: Blockchain · Tor · Anonymity · Unlinkable atomic swap

1 Introduction

Anonymity is an essential aspect of today's life in the digital arena. Through different mechanisms, every time someone visits a website, he/she can get tracked. In the wrong hands, such information can be used for malicious purposes, and the consequence of data leak can be devastating. For example, journalists around the world are dependent on anonymity while collecting evidence against powerful adversaries like a tyrannical Government. Societies and nations around the world depend heavily on the journalists' capability to hide their identities while collecting information. Journalism is a critical pillar of democracy, and without anonymity, the journalists would be at constant risk of loss of life.

Tor [1] is a technology that provides some reasonable anonymity to a web browser by using multiple independent relays to forward the network data. Tor or 'the onion routers' work in a way that no single relay knows the entire Tor path from the client to the server. Tor also gives the client the choice of the relays, which allows the client to choose relays from different countries to avoid getting victimized by a central authority.

However, currently, all Tor relays work voluntarily at their own cost. All classical channels of payment give up the payer's identity, thus making it impossible

J. Prieto et al. (Eds.): BLOCKCHAIN 2020, AISC 1238, pp. 42–51, 2020.
https://doi.org/10.1007/978-3-030-52535-4_5

for the client to pay for their service. While this works out for a small amount of traffic, it may not do so well if the demand for anonymity increases. One such problem with Tor is the downloading of large files. The voluntary unpaid Tor relays are forced to forward a large amount of data in such a case. The fact that the downloading of large files is a significant problem is documented in [2]. Also, the slowness of the Tor network has always been a problem, as described in [3]. Although this paper is relatively old, it is still relevant, as pointed out in [4]. Needless to say that the ability to get paid for the relay service would increase the number of relays, making it even easier to maintain anonymity.

From the beginning with Bitcoin [5], blockchains have always been considered an essential component for anonymous payments. Even though Bitcoin can be anonymous if used carefully, [6] shows that using Bitcoin inaccurately can cause deanonymization. There exist cryptocurrencies like Monero [7], for example, that provide even higher levels of security through ring confidential transactions [8]. However, such systems do not have smart contracts to this date, which is needed for our purpose. We consider a theoretical blockchain like Monero with some level of scripting support to be our blockchain of choice for implementing our idea. We describe the required scripting support in Sect. 3.

Since cryptocurrencies can facilitate anonymous payment, they make an obvious choice for paying the Tor relays for their service. In this paper, we describe how a fair payment mechanism can be achieved through a technique called an unlinkable atomic swap, which is also described in this paper. However, since cryptocurrency payments require some time to get finalized, our system can have a considerable amount of latency. Hence, we do not think this protocol is yet suitable for simple web-browsing. However, our protocol should be useful in case of downloading large files, in which case the latency would be tolerable.

2   Related Work

Since the problem of the lack of a sufficient number of Tor relays came to surface, there has been some research into paying the Tor relays. Good reviews have been published in [9,10]. We summarize the content here.

In the PAR system defined in [11], users buy coins from a bank which they use in the relay-requests to pay the Tor relays. The problem with this system is that there can be double spending where the same coin is used in multiple Tor requests. In the Gold Star scheme mentioned in [12], a central authority measures the bandwidth of each router by using randomized test packets sent to itself. The authority then rewards the fastest 7/8 of the relays with a Gold Star in a public listing. Other Tor relays can then prioritize the traffic coming from Gold Star relays, thus incentivizing the traffic that originated from the Gold Star relays themselves. This system only prioritizes Tor relays, so non-relay clients cannot take advantage of it. The BRAIDS system proposed in [13] solves the problem of double-spending by using relay specific tickets. However, just like the previous systems, it too suffers from relying on a central system distributing the tickets. Similarly, the LIRA system proposed in [14] also uses a central authority to manage incentives. LIRA requires that the clients that want to use a relay have to solve a probabilistic cryptographic puzzle to get access to the system, but the relays that provide bandwidth can learn winning solutions to these puzzles through the bank. It is constructed in a way so that the bank cannot know the identity of someone buying a winning solution. However, this system also suffers from being rewarded by a central authority.

A system called TEARS [15] is the first system that uses a decentralized cryptocurrency called Shallots to pay the relays. Shallots can be used to purchase relay specific priority passes of different values. It is believed that eventually, Shallots can be used to purchase other things as well, giving it the required value. A problem with TEARS is that it depends on a decentralized bandwidth measurement system to pay the relays, which has never been described. The Tor-Coin system mentioned in [16] uses a system similar to TEARS, which also uses a cryptocurrency to pay the relays. TorCoin allows the client to mine TorCoin if specific bandwidth requirements are met and then distribute it to the relays. However, TorCoin suffers from the possibility of a Sybil attack, since there is no mechanism to stop the same relay from pretending to be thousands of different relays to have a fair chance of being its client to then freely gift itself the mined TorCoins. [10] proposes a system in which the clients solve proof-of-work puzzles that can be used in mining pools for reward and then include that proof of work in the request for a priority bandwidth. This allows the relay to use the proof of work in the mining pool to claim the mining reward. A good thing about this approach is that it allows a market-driven approach to finding the right price for bandwidth without a central authority.

However, in any system that uses some kind of a priority ticket like PAR, BRAIDS, LIRA, TEARS and the proof of work system, a relay can choose not to correctly prioritize some of the clients even after receiving priority passes or tickets or proofs-of-work, because the bandwidth measurement only probabilistically measures the bandwidth. This calls for a fair payment system that ensures that the payment is not completed until the client receives the data, and the client cannot receive the data until it pays. There is also a problem of fixing the price of a priority token or tickets as there are no market-based systems to discover the price of the bandwidth.

3   Blockchain Model

We assume a blockchain with a UTxO (Unspent Transaction Output) model as used in Bitcoin [5] and Monero [7]. The UTxOs in our system supports some level of scripting like Bitcoin. However, due to anonymity concerns, we assume a transaction model like that in Monero with ring confidential transactions [8] to hide the true source of a transaction and the amount transferred in a transaction. We assume that the *init* method is invoked right after the UTxO is created, and can never be invoked again.

To spend the UTxO, one must call the *spend(spender, data)* method with appropriate arguments if the method exists. In such a case, a ring confidential signature must not be used so that the spender key and the challenge can be known.

This does not pose any risk to anonymity as the temporary public keys are generated in the same way as in Monero, and there are no balance payments, aka the transaction is single-input and single-output. If the method returns *true*, the transaction is valid; otherwise, it is invalid. We also assume that calling methods on spent transaction outputs have no effect. We also assume that the signature of the spender has been appropriately verified before the *spend* function is invoked.

4   Unlinkable Atomic Swap

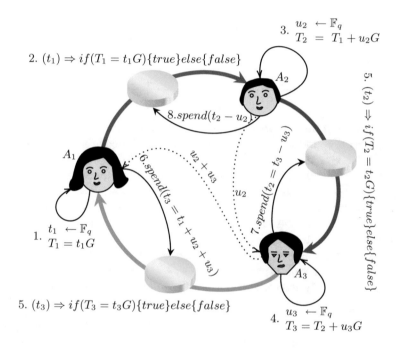

Fig. 1. Unlinkable Atomic Swap: solid lines represent transactions and dotted lines represent off-chain communication.

Atomic swap [17] is a well-known technique used to exchange cryptocurrencies in two different blockchains. The problem with an atomic swap is that anyone observing all the transactions can match the related transactions due to the same value of the challenge. When we are looking for strong anonymity, this is not acceptable. We suggest a variation of the atomic swap technique that prevents an onlooker from linking the related transactions. The unlinkable atomic swap also generalizes the atomic swap to a circular exchange among n parties. We assume that G is a generator in an elliptic curve group of prime order q so that the discrete log problem is hard on the curve. $\mathbb{F}_q$ is the field integer modulo q. We assume that $x \leftarrow \mathbb{F}_q$ means x is chosen randomly from a uniform distribution over $\mathbb{F}_q$. The protocol is described as follows and is also represented in Fig. 1.

Smart Contract 1: Conditional payment for unlinkable atomic swap

---

Globals: *expiry, C, payee, payer*

function *init(payer, payee, expiry, C)* :

| self.payee = payee; self.payer = payer; self.expiry = expiry; self.C = C;

end

function *spend(spender, c)* :

| **if** *spender==payer $\wedge$ currentTimestamp() > expiry* **then**

| | **return** true;

| **else if** *spender==payee $\wedge$ currentTimestamp() $\leq$ expiry $\wedge$ cG == C)*

| **then**

| | **return** true;

| **else**

| | **return** false;

| **end**

end

---

1. A_1 generates a random secret $t_1 \leftarrow \mathbb{F}_q$ and computes the challenge $T_1 = t_1 G$.
2. A_1 makes a conditional payment to A_2 with the condition to find t_1 such that $T_1 = t_1 G$ using $init(A_1, A_2, currentTimeStamp() + n\delta, T_1)$ in Smart Contract 1, where δ is the delay allowed per payee to redeem its transaction.
3. A_2 generates a random scalar $u_2 \leftarrow \mathbb{F}_q$ and computes $T_2 = u_2 G + T_1$.
4. Similarly, for all $2 < i \leq n$, A_i generates a random $u_i \leftarrow \mathbb{F}_q$. A_i then computes $T_i = u_i G + T_{i-1}$ in the order of their indices. This step for A_{i+1} is done after the next step is done for A_i.
5. A_i then makes a conditional payment to A_{i+1} for $1 < i < n$ using $init(A_i, A_{i+1}, currentTimeStamp() + (n - i + 1)\delta, T_i)$, and to A_1 for $i = n$ using $init(A_n, A_1, currentTimeStamp() + \delta, T_n)$. A_i then privately sends $s_i = \sum_{j=2}^{i} u_j$ to A_{i+1} for $i < n$ and A_n sends the value of $s_n = \sum_{j=2}^{n} u_j$ to A_1. It can be seen that $t_i = (t_1 + \sum_{j=2}^{i} u_j)$. Note that A_i can compute $\sum_{j=2}^{i} u_j$ by observing the value of $s_{i-1} = \sum_{j=2}^{i-1} u_j$ received from A_{i-1} and then computing $s_i = \sum_{j=2}^{i} u_j = u_i + s_{i-1}$.
6. A_1 can now spend the amount using $spend(A_1, t_n)$ in Smart Contract 1 since she knows t_1 and the sum $s_n = \sum_{j=2}^{n} u_j$ and can compute $t_n = (t_1 + \sum_{j=2}^{n} u_j)$.
7. The moment A_1 spends the amount, A_n can see the value of t_n. A_n computes $t_{n-1} = t_n - u_n$ and spend the UTxO paid to him/her using $spend(A_n, t_{n-1})$ in Smart Contract 1.
8. Similarly, A_i computes $t_{i-1} = t_i - u_i$ in reverse order to spend their UTxOs using $spend(A_i, t_{i-1})$. This completes the payment process.

It may be noted that the motivation for any of the nodes to start making the payment in the cycle may be different from the expectation of being able to unlock a payment from the previous node. Indeed, the secret held by the next node may be the key to something different. In our protocol for Tor relays, the secret is used to reconstruct the key to decrypt the data.

The unlinkable atomic swap has the following properties, assuming that a transaction emitted to a blockchain is always added to a block as long as it is valid -

Theorem 1 (Completeness). *If every party follows the protocol correctly, every party receives the correct payment.*

Proof. This is trivially true as the specific protocol is defined to work deterministically. □

Theorem 2 (Strong fairness). *Assuming all parties are probabilistic polynomial time algorithms, for any $1 \leq i \leq n$, if A_i's payment to the next party (A_{i+1} for $i < n$ and A_1 for $i = n$) succeeds but the payment to A_i from the previous party (A_{i-1} for $i > 1$ and A_n for $i = 1$) does not succeed, then A_i must have deviated from the protocol deliberately, even when all the other parties conspire together.*

Proof. Suppose, A_2 can spend its UTxO, but A_1 cannot. Since A_2 can spend its UTxO, that means A_2 could find the value of t_1. Since A_1 did not spend it's UTxO, the only details related to t_1 given up by A_1 is T_1. Hence, the parties A_2 to A_n together must have solved the discrete logarithm problem of finding t_1 given T_1, G such that $T_1 = t_1 G$ which has negligible probability.

Otherwise, for $2 \leq i \leq n$, while creating the UTxO for the next party, A_i needs to check that it has a conditional payment to itself with some challenge T_{i-1}. It must then create $T_i = u_i G + T_{i-1}$, where u_i is a random scalar chosen by A_i. If A_i did not daviate in this part of the protocol and if $T_i = t_i G$ and $T_{i-1} = t_{i-1} G$, then $t_i = u_i + t_{i-1}$. Now, since the payment to the next party succeeded, the value of t_i has been disclosed. This means, the value of $t_{i-1} = t_i - u_i$ can be computed by A_i, since u_i was generated by A_i. So, the only way A_i does not spend the UTxO from the previous party is if A_i does not spend it deliberately. □

Theorem 3 (Unlinkability). *Suppose the parties A_1 and B_1 are consecutive members of a cycle C_1, and A_2 and B_2 are consecutive members of cycle C_2. Then, a conditional payment between A_1 and B_1 in C_1 is indistinguishable from a conditional payment between A_2 and B_2 in C_2 to everyone other that A_1, B_1, A_2 and B_2 including any other party in either of the cycles.*

Proof. Since we use ring confidential transactions with the transaction amounts hidden by Pedersen commitments and the public keys are pseudorandom, we know that the only information that can possibly cause linkability is the challenge and the solution to the challenge. Let us use the index i for A_1 in C_1 and index j for A_2 in C_2, i and j not necessarily being equal. Let B_1 be the party next to A_1 in C_1 and B_2 be the party next to A_2 in C_2. Now, the challenge values are $T_i = T_{i-1} + u_i G$ and $T_j = T_{j-1} + u_j G$ in the corresponding chains. Now, $T_j = T_{j-1} + u_j G = T_{i-1} + (T_{j-1} - T_{i-1}) + u_j G$. Let $(T_{j-1} - T_{i-1}) = sG$ where s is the discrete logarithm of $(T_{j-1} - T_{i-1})$ with respect to G. Hence, $T_j = T_{i-1} + (s + u_j)G$. Since, $u_j \leftarrow \mathbb{F}_q$, we have $s + u_j \leftarrow \mathbb{F}_q$. Hence, both u_i and

$s + u_j$ have the same distribution. This implies T_i, T_j have the same distribution and hence cannot be distinguished. By a similar argument, it can be seen that the answers to the challenges in both the chains are also indistiguishable.   □

5   Design Objectives

We propose a modification of the Tor protocol that allows for securely paying the relays. We retain the following properties from the current Tor protocol -

1. No Tor relay can know both the source and the target of a request without colluding among one another.
2. Only the exit relay can perform a man-in-the-middle attack.
3. Only the exit relay can read the contents of the request and response.
4. Only the exit relay can modify the content of the request or the response.

The only property we do not retain is the low-latency property due to the latency of cryptocurrency payments. However, it may be tolerable in case of the transfer of a large volume of data in a single request.

In addition to the existing properties, we also provide the following additional properties -

5. The nodes can advertise a fee structure for their service.
6. The nodes are guaranteed that the client cannot access the response from the server without paying for their service.
7. The client is guaranteed to receive the correct data in case a successful payment of the fee as long as the exit node cooperates with the client.

6   Modified Tor Protocol

We modify the Tor protocol to enable payment to the relays. We use the technique of unlinkable atomic swap to enable this. The modified protocol works with the unlinkable atomic swap protocol for payment to the relays. The protocol is as follows.

1. Every Tor-relay is listed in a public list along with their public keys in servers called Tor directory services. There are three types of relays - guard relays, middle relays and exit relays
2. The client chooses one guard relay, one middle relay and one exit relay randomly from the pool.
3. The client builds the circuit in steps with Diffie-Hellman key exchange with each of the chosen relays. First the client uses Diffie-Hellman key exchange to establish a secret key k_1 with the guard relay. It then uses the guard relay to communicate with the middle relay to do a Diffie-Hellman key exchange with the middle relay to establish the secret key k_2. After that, the client uses the combined channel to establish the secret key k_3 with the exit relay.

4. Once the client establishes three independent shared keys k_1, k_2, k_3 with guard, middle and the exit relays respectively, the client encrypts the data D in layers. Let M_1, M_2, M_3 be the metadata (IP address, port number etc.) for the guard, middle and the exit relays respectively, and let M_D be the metadata of the destination. The client computes $E_3 = Enc_{k_3}(M_D||D), E_2 = Enc_{k_2}(M_3||E_3), E_1 = Enc_{k_1}(M_2||E_2)$, where the function Enc_k encrypts with the key k.

5. The client sends E_1 to the guard relay using the information in M_1. The client also sends a payment offer for the payment rate (per byte) I_1 to the guard relay along with the request.

6. If the guard relay likes the offer, it decrypts E_1 using k_1 to get M_2 and E_2. Otherwise it sends back a 'reject' message.

7. The guard relay sends E_2 to the middle relay using the metadata M_2. The guard relay makes a payment offer for rate $I_2 < I_1$ to the middle relay along with sending the request.

8. If the middle relay likes the offer, it decrypts E_2 using k_2 to get E_3 and M_3. It uses M_3 to send E_3 to the exit relay along with the payment offer for the rate $I_3 < I_2$. If it does not like the offer, it sends a 'reject' message to the guard relay.

9. If the exit relay likes the offer, it decrypts E_3 using k_3 to get D and M_D. The exit relay then delivers D to the server using the metadata M_D. If it does not like the offer, it sends a 'reject' message to the middle relay.

10. The server sends response R to the exit relay.

11. The exit relay generates a random scalar t_1 and computes $T_1 = t_1 G$. The relay then computes an encryption key $l = H(t_1)$ and computes $L = Enc_l(R)$

12. The exit relay computes $F_3 = Enc_{k_3}(T_1||L)$ and sends F_3 to the middle relay.

13. The middle relay sends $F_2 = Enc_{k_2}(F_3)$ to the guard relay.

14. The guard relay sends $F_1 = Enc_{k_1}(F_2)$ to the client.

15. The clients progressively decrypts F_1 with k_1, k_2, k_3 in order to get $(T_1||L)$.

16. The client computes the size s of the data L and creates a conditional payment of amount $s * I_1$ to the guard relay with the condition to find t_1 such that $T_1 = t_1 G$. The client does this by calling $init(Client, GuardRelay, currentTimeStamp() + 3\delta, T_1)$ in Smart Contract 1 where δ is the time delay, within which each relay has to spend the amount. The public keys $Client, GuardRelay$ are generated the same way as in Monero.

17. The guard relay generates a random scalar u_2 and computes $T_2 = u_2 G + T$. The guard relay then makes a conditional payment to the middle relay for amount $s * I_2$ with the condition to find t_2 such that $T_2 = t_2 G$ using $init(GuardRelay, MiddleRelay, currentTimeStamp() + 2\delta, T_2)$. The guard relay also sends u_2 to the middle relay.

18. The middle relay generates a random scalar u_3 and computes $T_3 = u_3 G + T_2$. The middle relay then makes a conditional payment to the exit relay for the amount $s * I_3$ with the condition to find t_3 such that $T_3 = t_3 G$ using

$init(MiddleRelay, ExitRelay, currentTimeStamp() + \delta, T_3)$. The middle relay also sends $u_2 + u_3$ to the exit relay.

19. The exit relay computes $t_3 = t_1 + u_2 + u_3$ and spends the amount using $spend(ExitRelay, t_3)$.
20. The middle relay computes $t_2 = t_3 - u_3$ and spends the amount paid to it using $spend(MiddleRelay, t_2)$.
21. The guard relay computes $t_1 = t_2 - u_2$ and spends the amount paid to it using $spend(GuardRelay, t_1)$.
22. The client now has access to t_1, and hence it can decrypt the content of R from L with the key $l = H(t_1)$.

The design objective 1 is ensured by Tor's design and the unlinkability of the payment design. The objectives 2–4 are maintained due to Tor's design. Objective 5 is ensured by the rate-listing of the relays on the directory services. And finally, the objective 6 and 7 are maintained due to the fairness property of the payment system.

It is still possible that they can still be linked by timing the transactions, which is possible in the regular Tor network as well under some circumstances. This can be avoided if there are plenty of transactions happening all the time.

7   Use with Tor Hidden Services

The Tor protocol documentation also has mentioned a way to have hidden services - i.e., servers with hidden location and network address, that are only accessible through some rendezvous points. The client first creates a circuit to the hidden service and then sends requests and receives responses from the server. Such hidden services are aware of the Tor protocol and can remove the need to trust such an exit relay to maintain the integrity of the data. The server itself can take the part of the exit node in terms of constructing the challenge that opens the encryption. Our protocol also allows the payment to the server along with paying the relays. Similar to the unlinkable atomic swap, our protocol can also be extended easily to more than three relays along with extending to the hidden services.

8   Summary

Our protocol provides a way to pay the Tor relays for their service while still maintaining anonymity. It also provides a guarantee to the Tor routers to get paid if the client can get the data, and on the other hand, it guarantees the client that it does receive the data as long as it pays the relays. The client does have to trust only the exit relay about the integrity of the data it receives, which is not different from the case of the present Tor protocol. However, in the case of Tor, using TLS prevents the exit relay from carrying out any of the attacks. That is true in our system as well. However, an exit relay in our case can get paid after serving garbage data, even though the client would figure it out after paying. Nevertheless, there is no obvious incentive for it to do so since it still has to stream an equal volume of data even in such cases. This problem does not exist in case of a Tor hidden service where the server knows about the protocol.

References

1. Dingledine, R., Mathewson, N., Syverson, P.: Tor: the second-generation onion router. https://svn-archive.torproject.org/svn/projects/design-paper/tor-design.pdf
2. Turning funding into more exit relays. https://blog.torproject.org/turning-funding-more-exit-relays
3. Dingledine, R., Murdoch, S.J.: Performance improvements on Tor or, why Tor is slow and what we're going to do about it (2009). https://svn-archive.torproject.org/svn/projects/roadmaps/2009-03-11-performance.pdf
4. Why is tor so slow? (2019). https://2019.www.torproject.org/docs/faq.html.en#whyslow
5. Nakamoto, S.: Bitcoin: a peer-to-peer electronic cash system. http://bitcoin.org/bitcoin.pdf
6. Jawaheri, H.A., Sabah, M.A., Boshmaf, Y., Erbad, A.: Deanonymizing tor hidden service users through Bitcoin transactions analysis. Comput. Secur. **89**, 101684 (2020)
7. Noether, S., Noether, S.: Monero is not that mysterious. https://web.getmonero.org/resources/research-lab/pubs/mrl-0003.pdf
8. Noether, S., Mackenzie, A., Monero Core Team: Ring confidential transactions. https://web.getmonero.org/resources/research-lab/pubs/mrl-0005.pdf
9. Tor incentives research roundup: GoldStar, PAR, BRAIDS, LIRA, TEARS, and TorCoin (2015). https://blog.torproject.org/tor-incentives-research-roundup-goldstar-par-braids-lira-tears-and-torcoin
10. Biryukov, A., Pustogarov, I.: Proof-of-work as anonymous micropayment: rewarding a Tor relay. In: Böhme, R., Okamoto, T. (eds.) Financial Cryptography. Springer, Heidelberg (2014)
11. Androulaki, E., Raykova, M., Srivatsan, S., Stavrou, A., Bellovin, S.M.: PAR: payment for anonymous routing. In: Borisov, N., Goldberg, I. (eds.) Privacy Enhancing Technologies, pp. 219–236. Springer, Heidelberg (2008)
12. Johnny Ngan, T.W., Dingledine, R., Wallach, D.S.: Building incentives into Tor. In: Sion, R. (ed.) Financial Cryptography and Data Security, pp. 238–256. Springer, Heidelberg (2010)
13. Jansen, R., Hopper, N., Kim, Y.: Recruiting new Tor relays with BRAIDS. In: CCS 2010 (2010)
14. Jansen, R., Johnson, A., Syverson, P.F.: LIRA: lightweight incentivized routing for anonymity. In: NDSS. (2013)
15. Jansen, R., Miller, A., Syverson, P.F., Ford, B.: From onions to shallots: rewarding Tor relays with Tears (2014)
16. Ghosh, M., Richardson, M., Ford, B., Jansen, R.: A TorPath to TorCoin: proof-of-bandwidth altcoins for compensating relays (2014)
17. Herlihy, M.: Atomic cross-chain swaps. CoRR abs/1801.09515 (2018)

A Blockchain-Based Approach for Cross-Ledger Reconciliation

Adriano Ribeiro[(✉)], Luiz Santos[(✉)], Alexandre Furtado[(✉)],
Bruna Schroder[(✉)], Daniel Vidaletti[(✉)], and Mariangela Vanzin[(✉)]

Eldorado Research Institute, Ipiranga Av, 6681, 99A,
Porto Alegre, RS 1301, Brazil
{Adriano.Ribeiro,Luiz.Santos,Alexandre.Furtado,
Bruna.Schroder,Daniel.Vidaletti}@eldorado.org.br, mvanzin@gmail.com

Abstract. Technologies available to implement cross-ledger reconciliation processes are centralized due to the strong necessity of consistency. These solutions fail to guarantee data integrity since data and business rules are handled separately, and the data is processed in batch, delaying the results. This article proposes a decentralized approach to the reconciliation processes using a private Blockchain based on the Ethereum platform. Consistency will be assured by using an Authority Round algorithm and the data will be processed in real-time, showing the results as soon as the transactions are processed. Since the records in the Blockchain are immutable, all the transactions will be traceable, allowing auditability and maintaining the link between original and generated data.

Keywords: ERP · Blockchain · Ethereum · Reconciliation process · Smart contract

1 Introduction

In large companies, the financial system [14] usually has a deep and complex ledger architecture. The account planning turns out to be distributed in multiple ledgers and reconciliation becomes intricate and expensive. Organizations like Oracle, SAP and Microsoft provide a solution called Enterprise Resource Planning (ERP). However, since costumers have different accounting processes, this solution needs to be customized before it can be used. If the ERP is the core process of a financial system, the reconciliation process is the heart of the company: it combines all journals generated by every ledger registered in the hierarchy. As companies expand their business, the transaction volume grows exponentially, thus making reconciliation a time-consuming problem.

This paper will discuss the usage of decentralization provided by Blockchain technology [9] using smart contracts [5] and Authority Round (AuRa) [3] consensus providing instant finality. AuRa is a Proof-of-Authority consensus therefore

J. Prieto et al. (Eds.): BLOCKCHAIN 2020, AISC 1238, pp. 52–60, 2020.
https://doi.org/10.1007/978-3-030-52535-4_6

it does not depend on nodes solving arbitrarily difficult mathematical problems, but instead uses a set of "authorities" - nodes that are explicitly allowed to create new blocks and secure the Blockchain. The chain has to be signed off by the majority of authorities, in which case it becomes a part of the permanent record. The goal of this work is to demonstrate that a decentralized approach to the cross-ledger reconciliation can improve reliability and performance, since data processing can be divided among several nodes without compromising consistency. The use of Blockchain enforces strong security and resiliency due to a combination of data and business rules in smart contracts. In addition, smart contracts can trigger events, that can be used as a light and fast notification interface to provide real-time communication with other systems.

The rest of this paper is organized as follows: Sect. 2 introduces the concept of cross-ledger reconciliation, discussing how it is implemented nowadays and the problems with this approach. Section 3 presents the proposed solution to solve those problems, including an experiment execution to show the implementation working in practice. Finally, Sect. 4 presents the experiment results and conclusions.

2   Context

2.1   Centralized Reconciliation

Cross-ledger reconciliation brings several challenges [18] since companies account planning are very specific, as well as hardware and software. There are multiple ways to organize the ERP coordination, however, in this work we will consider a company that uses Total Local Autonomy [18] to design your own account planning but needs to report all operation to a general ledger through an external cross-ledger solution.

Existing solutions implement ledger's reconciliation using centralized databases. The diagram below shows the information flow in this type of solution in Fig. 1, where the reader can see multiple companies generating journals (set of account transactions) and a consolidator that runs the reconciliation job to calculate the balance across different ledgers.

The journals come from multiple source systems as shown in Fig. 1, and each one have their own account planning structure, reconciliation rules and hierarchy. To do a reconciliation in this scenario it is necessary to have a set of transformation rules out of the company ledger to support cross-ledgers journal reconciliation. It is hard to guarantee a consistent data view of transactions and journals because of the constraints and characteristics of the technologies used to implement it. The data flow in this kind of solution would be something like: collect data from the source systems, load rules to be applied and run the transformations on data to generate new journals and transactions. This design makes it hard to implement data tracking because of the data separation business rules. Also, the current design has serious limitations about high data volume, mainly caused by centralization, and lacks transparency in the reconciliation process.

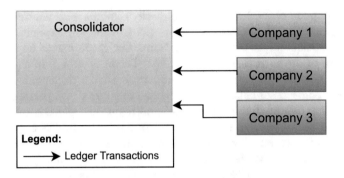

Fig. 1. Current process.

The scale of the reconciliation in terms of data volume is around 50.000 transactions per day. These transactions are collected from several source systems and processed multiple times in a day. The execution produces a report of the account state and shows the necessity of a full reconciliation journal's generation. Table 1 presents three days of collected data and executions.

Table 1. Total transactions processed per day.

Process execution date	Total records processed
09/03/2019	51137
09/04/2019	51361
09/05/2019	51116

During research very few decentralized approaches have been found for cross-ledger reconciliation, one of them being Spunta [20] a Blockchain based approach made by the Italian Bank Association. Spunta is a proof of concept of the usage of Blockchain for account reconciliation between 200 Italian banks. This experiment used a full year's data (200 million) from those banks.

2.2 Blockchain Decentralization

Ethereum is a worldwide network of interconnected computers that execute and validate programs. It provides a decentralized Turing-complete [15] platform called Ethereum Virtual Machine responsible to run application's code written as smart contracts. Several indexed data structures are provided by smart contracts to allow querying transactions and journals at any time, therefore Blockchain enables a real-time cross-ledger reconciliation solution as an alternative to the current implementation.

Blockchain has been used to track data and execute distributed logic [21] via smart contracts, as a security platform [13]. Also, it is agnostic to data format since all the data is stored in an efficient serialized format in a block. To load data into the Blockchain a smart contract was defined as an entry point to receive transactions from other systems. This smart contract executes all registered rules, transaction transformations and journal generation. All transactions loaded are versioned and every data derived is indexed and linked with the original data to enable the user to drill through the transactions.

Using the decentralization of the proposed solution it is possible to have a node for each company to send their own transactions, thus splitting the processing among multiple nodes. Table 2 displays three days of transactions grouped by company. Comparing these numbers to the ones processed by batch in Table 1, it is noticeable a downsize of data volume between 85% and 95% per entry point. The decentralization provided by Blockchain allows executing the reconciliation process and access to the current state of the account with just the normal data transversal.

Table 2. Transactions processed by company per day.

	Total records processed		
Process execution date	09/03/2019	09/04/2019	09/05/2019
Company A	4275	4189	5562
Company B	2741	6427	8581
Company C	2720	4128	5562

Decentralized smart contracts were used to generate real-time cross-ledger reconciliation state change notifications. To achieve these, the architecture proposed was based on Ethereum Blockchain and smart contracts to manage data linked with business rules [13].

The proposed solution focus on removing the necessity of a batch execution job to execute the reconciliation, as displayed in Fig. 2.

3   Experiment

3.1   Architecture

The architecture of the experiment was implemented with validator and non-validator nodes [17], each pair receives transactions from a different company. Different from the current centralized approach based on batch execution that strains the server, the Blockchain implementation process the transactions on demand and distribute it through multiple nodes. Therefore enabling to view partial account reconciliation in real-time. In Fig. 3 is presented an architecture generalization for a single company with two nodes: a validator and a non-validator. The non-validators nodes are responsible for receiving transactions from external applications and forward them to validators that will write those

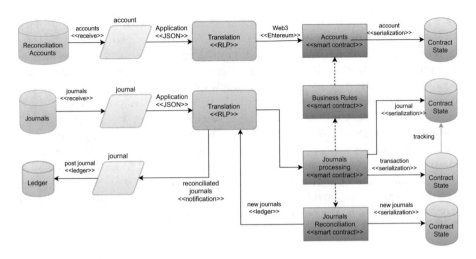

Fig. 2. Data flow of the proposed solution.

transactions into new blocks. Other responsibilities of non-validator nodes are event notification to subscribers and queries execution in smart contracts.

A RESTful API [7] was developed to be the proxy interface to the Blockchain. In the API, all the data received from external sources are converted to the Recursive Length Prefix (RLP) [4] serialization format to optimize data flow and storage in the Blockchain. All transactions need to be signed before being processed and replicated by the nodes. Thus, in order to improve performance, we use web3 framework [16] to sign the transactions in the API before sending them to the nodes. The API accesses encrypted files containing the validator accounts private keys, the keys are obtained from those files using the accounts passwords and the transactions are signed using Elliptic Curve Digital Signature Algorithm (ECDSA) [10].

Pentaho Data Integration (PDI) [6] is used to simulate the data flow to the Blockchain. At the beginning of the experiment thousands of transactions are loaded from a file and treated through PDI, these transactions are then sent

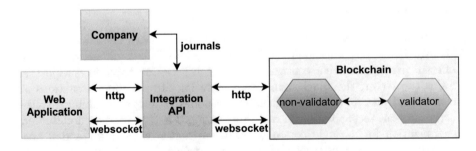

Fig. 3. Architecture of the solution.

to the Blockchain via API. The transactions are processed in smart contracts, which implement the cross-ledger reconciliation business rules.

There is a simple web application that allows the user to search transactions by date, and show the results in a grid. When looking at reconciliation transactions, a button can be pressed to show the original transactions that were accounted for. Every time a new journal is inserted, a WebSocket event is triggered by the Blockchain and a toast notification pops up on the user interface screen asking whether the user wants to reload the page.

To improve the performance, each Blockchain node runs on its own machine. The RESTful API and Pentaho Data Integration shares another machine. The operating system and hardware specifications are detailed in Table 3.

Table 3. System and hardware specifications

Number of machines	3
Operational system	Ubuntu Linux 18.04.3 LTS 64 bits
RAM capacity	16 GB
Hard disk	500 GB
Processor	Intel Core i7-4790 CPU @ 3.60 GHz x 8

3.2 Experiment Execution

The experiment consists of a reconciliation happening in real-time supported by a decentralized platform. Ethereum was selected for this project mostly because of the smart contract infrastructure [9] provided, that allow the development of decentralized applications. The decentralized nature of Ethereum allows a transparent data distribution of business rules and data as-is in every node of the Blockchain.

Three datasets were selected to be used in the experiment, one with the accounts to be transformed in the reconciliation process by the business rule engine and two journal datasets to simulate transactions flow from ledgers. The smart contracts architecture was organized taking in consideration their role: there are smart contracts to data translations, business rules, transformation logic and data storage. This design allows to update the smart contracts individually by its business context. In Ethereum data and logic work as a whole, therefore data migration is expensive in terms of Ether [17].

The data translation contracts work as the interface to the Blockchain, following the Proxy Pattern [11] implementation that reduces the Blockchain attack surface. In addition, those smart contracts do basic translations replacing some transaction field values to conciliate unbalanced journals.

A journal can have thousands of transactions, thus, to avoid sending big messages to the Blockchain and slow down the processing [2], the journal is partitioned in a reduced set of transactions before being sent to the smart contract that processes the transactions according to the specific journal. This contract has the logic to avoid losing track capability between journal and transaction.

Transformation logic receives individual transactions from data translation contracts, then calls the business rules engine and executes the reconciliation. This process generates events that could be used to notify other systems or user interfaces.

Business rules contracts are an inference engine proof of concept based on the indexed data structure to allow fast transverse and specific transformations execution. This data structure is composed by a set of predefined rules that filter the transactions that must be balanced. Balancing is achieved by creating a new transaction in order to zero the account balance.

Two types of business rules were implemented to validate the concept, first one is called "original" that identifies the source accounts of transactions and maintains original accounts in balance transaction; the other one, called "replacement", replaces the original accounts by another defined in the rule.

This architecture presents several advantages against the current one, mostly because it practically eliminates the necessity to load rules from the database before using them in the reconciliation process. Decentralization allows locally execution and eliminates the centralization requirement of the current implementation. The rules are replicated by the Blockchain every time they change, eliminating the necessity of a second service to synchronize.

A difference in this new implementation is that it eliminates the necessity of a job to start the reconciliation process. This proposal allows a direct connection to the ledgers enabling the processing of journals as soon they are received by the Blockchain as a natural data stream. Another difference is the possibility to easily track [13] transactions and journals. The tracking functionality is very important in accounting systems, and the proposed implementation supports this functionality by indexing and linking all transactions with their original data.

The last smart contract type is responsible to store all data sent to or generated by Blockchain. Store data in Ethereum is expensive and the EVM has a very limited stack, restricting the number of local variables in smart contracts methods. To overcome the contract update and data storage limitations in smart contracts, all transactions were stored in a serialized agnostic format called RLP. Thus, became possible to easily handle data inside Solidity functions reducing the necessity of local variables declaration. Using an agnostic data format eliminated data structure declarations for each different type of transaction.

The solution proposed has three main data flows, one to receive and persist data called write model providing strong consistency, another one representing a read model that facilitates filtering, indexing, etc. and the last one to synchronize both read and write. The implementation of these patterns with a platform like Ethereum amplifies their horizontal scalability because of the consensus layer.

4   Conclusion

The present work brings several contributions on the usage of decentralized architecture applied to cross-ledger reconciliation solutions. The decentralization helps dividing the workload since each company sends their transactions to a different node. The real-time processing characteristics of smart contracts allow replacing the reactive job scheduler by processing initiators that continually load data from source systems and processes data. The centralization requirement was eliminated because the consensus layer and smart contracts provide global ordering and strong consistency. The notification services interface was replaced by Ethereum subscribe interfaces with indexed logs.

A Blockchain-based architecture provides a flexible tracking infrastructure that brings data transparency in every point of the process. Through immutability, it guarantees an auditable process at any time. Despite the contributions, the proposed architecture has several areas of improvement in terms of data storage, rule management and peer-to-peer reconciliation. Storage could be improved with sharding [19] or using a controlled replication as implemented in Quorum [12] that uses privacy to hide transaction details from nodes that are not directly involved. Also, explore other data structure rules and replication models could improve rules execution and data management. The architecture allows evolving to a decentralized version of Command and Query Responsibility Segregation [1] and Event Sourcing [8] design patterns implementation.

References

1. Betts, D., Dominguez, J., Melnik, G., Simonazzi, F., Subramanian, M.: Exploring CQRS and event sourcing: a journey into high scalability, availability, and maintainability with Windows Azure. Microsoft Patterns & Practices (2013)
2. Bonér, J., Farley, D., Kuhn, R., Thompson, M.: The reactive manifesto (2014). http://www.reactivemanifesto.org/ [Dostopano: 21. 08. 2017]
3. Parity Wiki. https://wiki.parity.io/Aura
4. Chinchilla, C.: (2019). RLP https://github.com/ethereum/wiki/wiki/RLP
5. de Vilaca Burgos, A., de Oliveira Filho, J.D., Suares, M.V.C., de Almeida, R.S.: Distributed ledger technical research in Central Bank of Brazil. Technical report, Working Paper (2017)
6. Dixon, J.: Pentaho Open Source Business Intelligence Platform Technical White Paper. Pentaho Corporation, Orlando (2005)
7. Fielding, R.T., Taylor, R.N.: Architectural styles and the design of network-based software architectures, vol. 7. Doctoral dissertation, University of California, Irvine (2000)
8. Fowler, M.: Event sourcing, 18 December 2005
9. Blockchain based data provenance and integrity for secure IoT environments. In: Proceedings of the 1st Workshop on Blockchain-Enabled Networked Sensor Systems (BlockSys 2018), pp. 13–18. ACM, New York. http://doi.acm.org/10.1145/3282278.3282281
10. Johnson, D., Menezes, A., Vanstone, S.: The elliptic curve digital signature algorithm (ECDSA). Int. J. Inf. Secur. 1(1), 36–63 (2001). https://doi.org/10.1007/s102070100002

11. Izquierdo, M.A.J.: EIP 897: ERC delegate proxy (2018). https://eips.ethereum.org/EIPS/eip-897
12. Morgan, P.: Quorum Whitepaper. JP Morgan Chase, New York (2016)
13. Approach for data accountability and provenance tracking. In: Proceedings of the 12th International Conference on Availability, Reliability and Security (ARES 2017), p. 10. ACM, New York. http://doi.acm.org/10.1145/3098954.3098958
14. Tuarob, S., Strong, R., Chandra, A., Tucker, C.S.: Discovering discontinuity in big financial transaction data. ACM Trans. Manag. Inf. Syst. **9**(1), 26 (2018). http://doi.acm.org/10.1145/3159445
15. Turin, A.M.: On computable numbers, with an application to the entscheidungs problem. Proc. London Math. Soc. **s2–42**(1), 230–265 (1937). http://doi.acm.org/10.1112/plms/s2-42.1.230
16. Vogelsteller, F.: (2015). web3. js. https://github.com/ethereum/web3.js
17. Wood, G., et al.: Ethereum: a secure decentralised generalised transaction ledger. Ethereum project yellow paper, vol. 151, pp. 1–32 (2014)
18. Markus, M.L., Tanis, C., Fenema, P.C.V.: Multisite enterprise resource planning implementation. Commun. ACM **43**(4), 42–46 (2000). http://doi.acm.org/10.1145/332051.332068
19. Faq, S.: https://github.com/ethereum/wiki/wiki/Sharding-FAQ
20. Fintech, Banks: Spunta the sector blockchain passes test of annual data. https://www.abi.it/DOC
21. Sun, H., Huang, Y.: Distributed ledger technology and economic contract innovation. In: Proceedings of the 3rd International Conference on Crowd Science and Engineering (2018). Article 15. https://dl.acm.org/doi/abs/10.1145/3265689.3267929

Towards a Secure Data Exchange in IIoT

Anna Sukiasyan$^{(\boxtimes)}$, Hasmik Badikyan, Tiago Pedrosa, and Paulo Leitão

Research Centre in Digitalization and Intelligent Robotics (CeDRI), Instituto
Politécnico de Bragança, Campus de Santa Apolónia, 5300-253 Bragança, Portugal
an.sukiasyan@gmail.com, {h.badikyan,pedrosa,pleitao}@ipb.pt

Abstract. Industrial Internet of Things (IIoT) plays a central role in
the Fourth Industrial Revolution, with many specialists working towards
implementing large scalable, reliable and secure industrial environments. However, existing environments are lacking security standards
and have limited resources per component which results in various security breaches, e.g., trust in between the components, partner factories or
remote control units with the system. Due to the resilience and its security properties, combining blockchain-based solutions with IIoT environments is gaining popularity. Despite that, chain-structured classic
blockchain solutions are extremely resource-intensive and are not suitable for power-constrained IoT devices. To mitigate the referred security
challenges, a secure architecture is proposed by using a DAG-structured
asynchronous blockchain that can provide system security and transactions efficiency at the same time. Use-cases and sequence diagrams were
created to model the solution. The achieved results are robust, supported
by an extensive security evaluation, which foster future developments
over the proposed architecture. Therefore, as the proposed architecture
is generic and flexible, its deployment in diverse customized industrial
environments and scenarios, as well as the incorporation of future hardware and software, is possible.

Keywords: Blockchain · IoT · Industrial IoT · Cybersecurity

1 Introduction

The number of companies approaching the Industry 4.0 paradigm is growing on
daily bases. Companies are connecting their devices to the internet to increase
the system's productivity and efficiency. In these Internet-connected environments, the security issues are the most challenging aspects to deal with. According to Cisco Annual Cybersecurity Reports, 31% of companies have experienced
attacks on Operational Technologies (OT) [1]. Despite the fact that 75% of
experts think of security as a high priority component, only 16% are sure that
the company is prepared to face the cybersecurity issues. The main reason for
that is the lack of standards for Industrial Internet of Things (IIoT) environments, endpoints and communication protocols.

J. Prieto et al. (Eds.): BLOCKCHAIN 2020, AISC 1238, pp. 61–70, 2020.
https://doi.org/10.1007/978-3-030-52535-4_7

The fourth industrial revolution includes several segments such as logistics and supply chain, transportation, mining, healthcare, oil and gas. The digital transformations are implemented with the use of Information and Communication technologies (ICT), Internet of Things (IoT), artificial intelligence, robotics, smart decentralized manufacturing infrastructures and self-optimizing systems in information-driven, cyber-physical environments. In the industrial world, Cyber-Physical Systems (CPS) can be seen as Industrial Control Systems (ICS), which can ensure that technical facilities run automatically by controlling business processes. ICS usually comprise Supervisory Control and Data Acquisition (SCADA), Distributed Control System (DCS), Programmable Logic Controller (PLC), Remote Terminal Unit (RTU), Intelligent Electronic Device (IED) and the interface which is to ensure the communication of components. Systems mentioned above are the building blocks of the critical infrastructures, meaning that reliability, availability and privacy of those systems are the main concerns. The protection of an IIoT system or the system state can be achieved by establishing and maintaining the system in a way to prevent unauthorized access to the system or its resources. This will also prevent data loss or major damage in the system. ICS were usually isolated systems using proprietary control protocols. Nowadays, as IT solutions are being integrated into ICS environments, they are becoming open for remote access and working on improving connectivity between system components. There are various standards and solutions for IT environments security, but those can not be applied to ICS due to several specific requirements [2]:

- Functional requirements: ICS as part of production processes have many components of the system that are embedded, which reduces the possibility of classic security solutions being directly applicable to the production. From the production perspective, confidentiality is the main risk, but availability still stays the first priority in the requirements list.
- Resource requirements: Many ICS are running on real-time operating systems which is a highly resource consuming process. Also the components of the ICS normally have low processing power and machine specific limitations that reduce the chance of being able to perform security updates on the system components.
- Security requirements: Industrial systems can contain confidential information about the production processes or the industry components. Loss of this information can result in the violation of company confidentiality or data leaks related to the production environment topology. These data leaks will not result in the loss of equipment, but can be used in future attacks.

Having this in mind, this paper explores the current state of the art of the security in IIoT environments by identifying the potential threats and the current capability of devices enrolled in the industrial environments, and purposes a solution to enable secure data exchange in IIoT.

The rest of the paper is organized as follows. Section 2 presents a characterization of the security issues in IIoT environments and Sect. 3 overviews the use of blockchain in IIoT. Section 4 describes the proposed secure data exchange approach for IIoT environments, and finally, Sect. 5 rounds up the paper with the conclusions and points out the future work.

2   Security Characterization of Industrial IoT

IIoT security surveys show that IIoT endpoints are the main source of system's vulnerabilities, their definition being dependent on the system architecture, i.e. an endpoint can mean the IoT device itself or a group of devices that are responsible for any particular operation or performing any role in the system. Endpoints are managed through the network and are used for the data exchange, data collection or control purposes. Around 72% of the endpoints rely on the use of Internet protocols and 53% are IP-based, domain-specific protocols that are replacing point-to-point, non-routable protocols for control systems. The most commonly used protocols are MQTT (Message Queuing Telemetry Transport) and CoAP (Constrained Application Protocol) as they overcome others in terms of header size, power consumption, and data loss [3].

The ICS architecture consists of 2 layers: physical layer that includes all sensors and hardware components, and cyber-layer composed mainly from SCADA systems. SCADA systems are a set of protocols, platforms and technologies used to manage an ICS. Traditionally, the protection of SCADA systems has been based on the physical isolation, using non-standard protocols. The components responsible for the communication between other services are a direct target for attacks that can be solved by using secure network protocols covering authentication, confidentiality and integrity aspects. But in industrial automation, similar protocols are hard to find, so the main priority is meeting the real-time requirements [4].

The use of secure protocols and intermediate pre-checks leads to performance issues and communication delays in time-critical infrastructures, so it is crucial to find the balance between latency and security. The interaction of communication components with external networks implies the importance of protecting transmitted data, as well as the access to communication functions. Network interconnection points, e.g., wireless access points, are also intrusion points and need to be monitored by Intrusion Detection System (IDS). For sharing information in external and internal networks, additional routers and firewalls are being deployed by IDS, which are capable of identity checks and traffic analysis. Similar solutions are used to protect the gateways [5].

3   Blockchain in IoT

Blockchain based systems are classical distributed systems that can be classified into two main types: permissionless and permissioned. Permissionless systems are publicly open for use while permissioned platforms are designed in a close-ended manner. This means that the permissioned system has a well defined and fixed set of nodes [6].

The features of the blockain's decentralized consensus system may be integrated with IIoT environments to mitigate some security issues. Most of the existing solutions are adopting chain-structured blockchain in IoT systems. This type of blockchain can bring limitations related with the consensus model as it can collide with the requirements in IoT field, such as low latency and high performance. Three main challenges of integrating IoT with blockchain are:

1. The trade-off between efficiency and security.
2. The coexistence of transparency and privacy.
3. The conflicts between high concurrency and low throughput.

Based on the referred challenges, the blockchain development is evolving into different variations of the classical idea, which according to the differences in the structure can be classified as:

Chain-Structured Blockchain: In chain-structured blockchain systems (Fig. 1), the longest chain of blocks is considered as the main chain for the system. If more then one block has been generated at the same time, the first generated block will join the main chain and for the other blocks, there a fork will be created. Only transactions placed in the main chain will be considered valid, which means that all transactions in secondary chains will be labeled as invalid blocks. Mechanisms implemented in traditional blockchains such as ZK-snark and the AZTEC protocol now used in the Ethereum are creating a highly secure environment, but at the same time elliptic curve arithmetic operations required by the AZTEC protocol are highly resource intensive [7]. Overall, chain-structured blockchain solutions are not suitable for power-constrained IIoT environments, where most of the components have low processing power and all transactions are performed in a time critical manner.

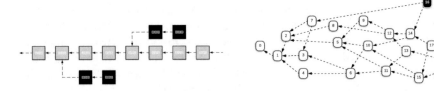

Fig. 1. Chain-structured blockchain architecture diagram

Fig. 2. DAG-structured blockchain diagram

DAG-Structured Blockchain: Aiming to integrate blockchain with more critical environments such as IoT, a new structure of blockchain has been created based on acyclic graph architecture, which is called tangle. In tangle, the concept of blocks is changed to an individual node representing each transaction in the distributed ledger. Unlike the first blockchain, the tangle uses different approaches to improve the throughput of the system which is a critical metric in the IIoT environment. It adopts asynchronous consensus model and as shown in the Fig. 2, the network is not limited to one main chain. It forks all the time by forming a tangle net. There are several implementations of DAG-structured blockchain, such as IOTA, ByteBall and NANO [8].

4 Proposal for Secure Data Exchange in IIoT

This section presents a solution for increasing security in IIoT environments by using the blockchain technology. The proposed approach considers a DAG-Structured blockchain security solution implemented on top of existing components in IIoT architectures. Due to the specifications of IIoT environments, which are time and resource critical, these requirements have been taken into consideration during the designing of the solution. The solution consists of 2 main parts: access control and secure transaction chain generation to ensure trust and data consistency in the system. As previously discussed, nodes of the industrial environment may have limited resources and can be divided into 2 types based on their processing power capabilities: light nodes and full nodes. In our solution only full nodes, such as gateways and managers, are considered members of a tangle network. The light nodes are connecting to the full nodes to publish a transaction to the network. The full node will sign each transaction received from a light node on their behalf, if the light node doesn't have this capability, and will publish it to the tangle network by using the IRI interface (IRI is an implementation of IOTA that also provides HTTP REST interface, so that light nodes can send transactions to the full nodes).

4.1 Architecture

Figure 3 depicts the architecture that will support the proposed solution. The architecture is composed by diverse components, namely the wireless devices, gateways, managers and the tangle network.

Wireless Devices: Wireless devices can be of the main 3 types: sensors, actuators and controllers. In IIoT environments, wireless devices are categorized as light nodes as they have limited resources and are not capable of using secure protocols or performing any power-consuming actions. Each device needs to have a unique identifier in the system and has to pass the authentication every time when trying to perform a transaction. As light nodes do not have enough processing power to implement Proof of Work (POW), they are not considered to be a direct part of the network. Light nodes will be able to send transactions to the network through the middleware which will serve as a gateway. During the registration process, each device in the system will be granted with a public/private key pair that will be used in future for signing transactions. The key pair generation will be performed by the gateway.

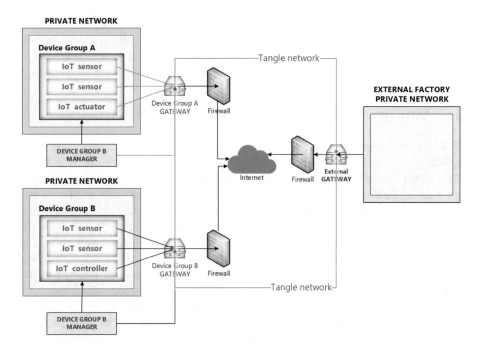

Fig. 3. Architecture diagram of the proposed solution

Gateways: Gateways serve as a secure middleware in between light nodes and the tangle network. As gateways are considered as full nodes, they are responsible for the tangle network maintenance. They also perform a role of a checkpoint which only submits transactions from the light nodes that are authorized by the manager. Gateways can be of 2 types: device gateway and external gateway. The first one is responsible for the key generation, authentication of group of devices (light nodes) and organization of the communication on their behalf. It also has capability to translate commonly used protocols to HTTP to deliver messages from device to the http endpoint of the tangle network. External gateways are responsible for the communication between 2 factories. They are the first access point for all the requests incoming to our industrial infrastructure from the outside. Gateways are the core components of the architecture that need to be set up in order to be able to start devices registration and communication processes in the system.

Manager: The manager is also a full node that is responsible for the device management in the system. The registration of the IoT device in the system is performed manually by the system administrator. After the device enters in the system, it will be registered in the device list by the manager, which is the only one that has permission to write for the device list. Other full nodes of the system only have read permission for the device list. These access control rules are also designed to increase the security in the system by preventing third party devices from making unauthorized changes. As mentioned above, devices will be divided by device groups. There is a limitation to have one manager node per device group. Manager has to be predefined and set up before being able to start the registration process for the light nodes.

Tangle Network: Tangle network in our architecture is a public blockchain network which allows any parties to participate in the process. It serves as the main solution for the trust issue in the system and allows us to have a consensus in the system for all published transactions. This is a requirement in order to be able to perform transactions between different industrial environments or remote nodes of the system regardless of their geo-location and security implemented on each individual device. The tangle network structure allows protecting system against several attacks, such as Distributed Denial of Service (DDoS), double-spending, etc. It also improves throughput of time and resource critical environment in comparison to chain-structured blockchain.

4.2   Functionalities

The functionalities provided by the proposed solution are the registration of devices, revoking devices, disable/restore devices and communion between 2 devices from different devices groups [9]. This paper focuses on the description of the communication between the devices as this is the scenario where most of the attacks are identified and should be mitigated by the proposed solution.

The communication between the devices that belong to different device groups is organized through the device group gateways, with 4 main components participating in this process: source and destination devices and their gateways. As mentioned earlier in the architecture diagram Fig. 3, the communication is performed through the tangle network. The source device will generate the package that need to be delivered to the destination. In the destination of the package, both gateway and device need to be specified. The package is sent by the source device to the device group gateway. Normally, as sensors are using industrial protocols for communication, the package will be passed to the translation module of the gateway. After being translated from industrial protocols to HTTP, the gateway is submitting the package as a transaction to the tangle network on behalf of the source device. After the transaction is approved on the tangle network by other nodes, the destination device group gateway will be notified about a new transaction in the network, as all the gateways are full nodes on the tangle network. As soon as the gateway will get the notification about the published transaction, it will convert the package from HTTP to industrial protocol appropriate for the destination device. After the translation, the package will be sent to the destination device.

More detailed actions performed during the communication process are shown on the sequence diagram represented in Fig. 4.

The sequence diagram illustrated in Fig. 4 is showing the steps performed to deliver data from device A to device B. The tangle network is shown as a

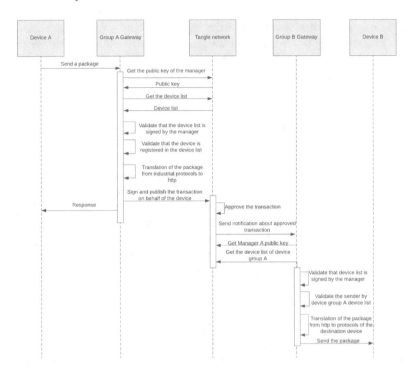

Fig. 4. Sequence diagram: communication between 2 devices from different device groups

separate node on the diagram but in the actual implementation all gateways will be published to the tangle network as full nodes, so the network will not be a standalone component of the system.

This architecture is flexible enough to allow removing the device group gateways from the current position and organizing direct communication between devices by using the tangle network in the future when the devices will have required processing power to be able to handle all the processes of the workflow described above. The lack of standards for IIoT devices and communication protocols brings challenges that can be addressed by the implementation of translation modules on the gateways. The semantic gateways can serve as a translator between the various communication protocols to allow the industrial environment growth independent from vendor-specific implementations.

4.3   Bootstrapping the System

The IOTA technology is used for the implementation of the tangled network, specifically the approach to create a private network, which allows to isolate the network and keep it accessible only for the nodes in our environment. Also, the current architecture allows to have a shared private network between multiple factories or industrial environments that will serve as a communication method in between them.

All components are set up and running on docker containers. For bootstrapping, a private tangle network, the components need to be set up and configured in the following order. First deploy of the Coordinator (COO), then start the IRI node and then launch the Coordinator service. At last, configure the node to subscribe the events in order to be notified when such are sent to the network.

After having the tangle network setup and running, device group gateways need to perform their first transactions in the network. The first transaction performed by the manager will be publishing its public key to the tangle, and the first transaction performed by the gateway is reading and storing the service group manager's published public key and storing it in the cache in order to be able to do the verification checks during the future communications. If for some reason, the manager will change or the key pair will be regenerated, a new public key will be published by the manager and all the nodes with already cached public key will be notified about the changes. On the other hand, the first transaction of all full nodes in the device group except for the manager is read request for the public key of the manager.

After performing this bootstrapping sequence, the system will be fully functional and all previous presented functionalities will be ready to use.

5   Conclusion and Future Work

This paper analyses how blockchain technology can be used to improve secure data exchange in IIoT, addressing specific requirements, such as time and resource critical that have an impact in the type of consensus that can be used on the blockchain.

The proposed solution is based in 2 logical groups: light nodes and full nodes. Light nodes are considered to be the ones that don't have the capability to implement any security functions, communicate via secure protocols or participate in the transaction approval and proof of work processes on the tangle. Full nodes participate in all processes, in the tangle and in the industrial environment and are also responsible for publishing transactions to the tangle network in behalf of the light nodes. Public/private keys are generated for each component of the system that are used for authentication and authorization purposes. The designed architecture provides a solid ground for trust assurance between all industrial components, by also providing a secure communication channels for remote control and data exchange. STRIDE threat analysis performed has shown that most of the attack vectors falling into the scope of the mitigation mechanisms presented are covered in the designed solution.

For future work is considered the development of the proposed architecture. After will follow the test in industrial environment replicating a real world scenario, to check the usability of the solution. Performance analysis should be done and optimization of various processes might be required because industrial environments are highly time and resource critical. One of the risks related to the performance that can arise is due to the growing chain of transactions in the tangle network. Growth of the transaction chain can increase decision making time for the approval of the transactions by all the nodes participating in the consensus. With the continuous monitoring of the implemented solution we need to make sure that no perceptible downgrade of the performance is identified.

Overall, the tangle network is a growing project used in various IoT based environments. Every day devices and sensors enrolled in the industrial systems are gaining more processing power and becoming capable of performing more complex calculations. Some security related functions will start to be made based on the light nodes, which will improve the trust and security. Probably some of the light nodes will gain capabilities to turn into full nodes and will participate in all processes equally. Our architecture is designed in a way to be agnostic to that future use case scenario. That means that the architecture is flexible enough to easily adjust to the predictable nearest future.

Acknowledgements. This work has been supported by FCT – Fundação para a Ciência e Tecnologia within the Project Scope: UIDB/05757/2020.

References

1. Cisco. Annual cybersecurity report (2018)
2. Fan, X., Fan, K., Wang, Y., Zhou, R.: Overview of cyber-security of industrial control system. In: Proceedings of the 2015 International Conference on Cyber Security of Smart Cities, Industrial Control System and Communications, SSIC 2015, pp. 1–7 (2015). https://doi.org/10.1109/SSIC.2015.7245324
3. Frustaci, M., Pace, P., Aloi, G., Fortino, G.: Evaluating critical security issues of the IoT world: present and future challenges. IEEE IoT J. **5**(4), 2483–2495 (2018). ISSN: 23274662
4. Neumann, P.: Communication in industrial automation-what is going on? Control Eng. Pract. **15**(11), 1332–1347 (2007). https://doi.org/10.1016/j.conengprac.2006.10.004. ISSN: 09670661
5. Hong, S., Lee, M.: Challenges and direction toward secure communication in the SCADA system. In: CNSR 2010 Proceedings - 8th Annual Conference on Communication Networks and Services Research (2010)
6. Baliga, A.: Understanding blockchain consensus models. Whitepaper, no. April, pp. 1–14 (2017). https://www.persistent.com/wp-content/uploads/2017/04/WP-Understanding-Blockchain-Consensus-Models.pdf
7. Williamson, Z.J.: The AZTEC protocol. Whitepaper, pp. 1–24 (2018)
8. Huang, J., Kong, L., Chen, G., Wu, M.-Y., Liu, X., Zeng, P.: Towards secure industrial IoT: blockchain system with credit-based consensus mechanism. IEEE Trans. Ind. Inf. **15**(6), 3680–3689 (2019)
9. Sukiasyan, A.: Secure data exchange in IIoT. Master thesis in Information Systems – Polytechnic Institute of Bragança (2019)

An Architecture for Sharing Cyber-Intelligence Based on Blockchain

Rui Gonçalo[1]([⊠]), Tiago Pedrosa[2], and Rui Pedro Lopes[2]

[1] Polytechnic Institute of Bragança, Bragança, Portugal
rgoncalo@ipb.pt
[2] Research Centre in Digitalization and Intelligent Robotics (CeDRI),
Instituto Politécnico de Bragança, Bragança, Portugal
{pedrosa,rlopes}@ipb.pt

Abstract. Cyber-intelligence sharing can leverage the development and deployment of security plans and teams within organizations, making infrastructures resilient and resistant to cyberattacks.

To be efficient, information sharing should be performed in a trusted environment, ensuring both the integrity, privacy and confidentiality and the truthfulness and usefulness of the information. This paper addresses this issue with the development and deployment of an architecture based on blockchain technology. Each participant is granted a reputation level, that is used to assess and verify the information other actors produce. Each actor, then, is given an amount of credit, corresponding to the number and accuracy of the validation. Information is also organized in topics, instantiated in independent ledgers. The architecture was validated with a three organization scenario, for proof-of-concept.

Keywords: Blockchain · Hyperledger Fabric · Cybersecurity · Intelligence sharing

1 Introduction

The volume of cyberattacks to organizations grows exponentially everyday, caused not only by the increase of the hackers' creativity but also because some organizations lack the concern or the resources needed to raise proper defences. When it comes to cybersecurity, while the cybersecurity engineer has to plug every leak or vulnerability, the hacker needs only one successful exploit to steal data or to disable a system [1]. One of the key points in cybersecurity is the complete knowledge of the full organization's Information Technology (IT) infrastructure, including the topology, equipment, operating systems and applications, as well as the collection and analysis results of artifacts from network incidents, or from suspicious activity records [2]. These should be used as the starting point for a complete understanding and development of a threat model, used to prioritize the interventions and configuration of the infrastructure [3].

J. Prieto et al. (Eds.): BLOCKCHAIN 2020, AISC 1238, pp. 71–80, 2020.
https://doi.org/10.1007/978-3-030-52535-4_8

The gathered intelligence can also be useful for future reference and it can be leveraged by others, if shared properly and trustfully.

This paper proposes an environment where communicating entities enter and exchange information with an initial reputation level right from the beginning, with this level increasing with time. The use case for the architecture described herein is the collection and share of cyber-intelligence, used to leverage the organization's security programs, thus encouraging each participant to contribute.

Trust levels are associated with tiers in the infrastructure and each participant is placed in a tier according to the credit it gathers. Each participant is thus classified according to its reputation: the more relevant the intelligence it shares, the more it increases in reputation and, consequently, move up in the tier system. The infrastructure is built on a permissioned blockchain platform, to securely record transactions, and exchange information between one another [4]. The blockchain platform is based on Hyperledger Fabric, a modular and extensible open-source system for deploying and operating permissioned blockchains [5].

The rest of the paper is organized as follows. Section 2 discusses blockchain technology and the consensus model. Section 3 describes the architecture and implementation details. Section 4 presents the test scenario and Sect. 5 the conclusions and discusses future work.

2   Background

Blockchain technology has been, since it was announced by Satoshi Nakamoto, in 2008 [6], one of the mainstream scientific and technological research topic in many areas, such as financial, healthcare, insurance, internet of things and many others [7].

In its essence, blockchain is a distributed database that records transactions between parties or digital events that have been executed and shared in the network [8]. It is usually represented as a chain of blocks intrinsically connected to each other (Fig. 1).

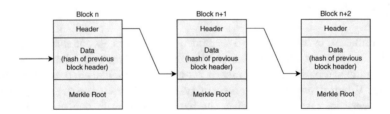

Fig. 1. Representation of a blockchain

Each block is cryptographically linked to the previous one, after validated by a consensus decision. As the chain grow, by the addition of new blocks, older blocks become more difficult to modify. Each block also stores a set of records, used to store transactions in a shared ledger. After a successful transaction, the block is added and replicated through the network.

2.1 Consensus Models

As mentioned before, the blockchain grows after consensus, which requires a broad agreement between several processes, even in the presence of faults or malicious nodes. This, known as the Byzantine generals dilemma, is characterized by the lack of trust between the participants. This dilemma describes the scenario of the lack of agreement in a group of generals, as to whether to attack or retreat from a siege to a city, and if there is no consensus it all falls apart [9].

Achieving consensus in a distributed system is challenging. Algorithms have to be resilient to failures of nodes, partitioning of the network, message delays, out-of-order and corrupted messages [10].

The research on blockchain technology introduced several consensus models, an essential part of the operation as they ensure consistency and availability of the ledger. One of the most well-known consensus model is called Proof-of-Work (PoW), that consists in using computing power to calculate a hash value that is less than a certain number, referred to as the difficulty level, and thus execute transactions over the network, rewarding a specific individual with the privilege of inserting the next block of transactions in the chain.

This model requires each node to use a considerable computation power, and a few alternatives appeared. One example is consensus by Proof-of-Stake (PoS). This model dictates that the network participants who have a portion or stake of the network value, such as digital coins, receive a proportional allowed mining power, implying that participants with more stake would less likely harm or dominate the network because the cost needed would not be profitable over normal mining activities [9].

Another alternative is the Raft consensus algorithm. Raft implements consensus by first electing a distinguished leader, then giving it complete responsibility for managing the replicated log [11,12]. The existence of a leader simplifies the algorithm. If it fails or become disconnected from the other servers, a new leader has to be elected. Raft, thus, assumes three steps for consensus: leader election (a new leader must be chosen when an existing leader fails), log replication (the leader accepts log entries from clients and replicate them), and safety (if any server has a particular log entry in its state machine, then no other server may apply a different command for the same log index).

2.2 Smart Contracts

Smart contracts, or decentralized applications, are programs that are executed on the blockchain, in which their correct execution is enforced by the consensus model. This includes any set of rules that can be represented in the programming language. The executed code, although generic and able to be used in different areas and with different purposes, can assume the role of the middle man, verifying certain criteria before executing a transaction, therefore eliminating negotiation hindrances, such as having a third party to verify if a transaction is valid. This smart contracts are triggered upon an event in the network, which means that they are self-aware upon events and automatically execute an action, written with rules agreed before hand by all participants, resulting in an practically unbreakable contract [7].

The implementation of permissioned blockchains usually requires to assemble a set of technologies in a common platform. For that, existing platforms and implementations may be used, saving the burden of implementing all the details of a blockchain operation.

2.3 Platforms and Implementations

There are several factors to consider in the selection of the best blockchain platform, such as development maturity, programming languages, consensus model, popularity and support, scalability, and smart contracts support.

One of the most popular platforms is the Ethereum platform and consists in a peer-to-peer network of virtual machines, that can be used to deploy and develop applications, although, it is more comfortable to support application programs based on rules and criteria [13]. Ethereum currently uses the PoW consensus model to reward the mining activity to the network participants. Furthermore, Ethereum also allows anyone to develop applications within the platform, as a permissionless system.

Along with permissionless blockchain platforms there are also permissioned blockchain implementations. One of this implementations is Ripple, which is a payment dedicated network that makes currency transactions rapidly. Ripple operates under the Ripple Protocol Consensus Algorithm (RPCA). This means that nodes have a list on which other nodes they trust rather than to accept any global assumption.

Another permissioned blockchain platform is the Hyperledger Fabric. This platform is modular since it allows for pluggable consensus models, although, raft is currently the default option. Usually, in the distributed ledger technology, consensus is considered to be a single algorithm applied to one part of the transaction flow. However, in Fabric, the consensus concept comprehends not only the algorithm itself but also most of the transaction process, from the transaction proposal until the transaction is verified and submitted into the ledger.

3   An Architecture for Sharing Information

The objective of this paper is to present an architecture for the exchange of reliable and meaningful intelligence in a partially trusted environment. A permissioned blockchain will be used as a communication mechanism to secure content and reward the participants that share information by increasing its reputation level.

The exchange of information relies on the concept of a channel, where each channel is specific for a context or topic. This is similar to message queues, although the content is not removed when read. Moreover, in addition to support the communication of content, the system also supports the validation of the content, performed by trustful peers (Fig. 2).

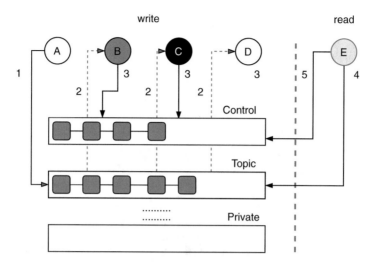

Fig. 2. Channel architecture flow

Although this approach relies on some tools and mechanisms used in some studies and articles [14], what makes it diverge from the rest is that its essence is based on a tier system. In the beginning of the network all organizations assume the same basic level, from which they are able to climb, by submitting reliable intelligence.

Each organization is assigned a reputation level, among five possibilities: *Others*, *Vendors*, *Sec. Researchers*, *CSIRT*, and *ICSIRT* (International CSIRT). The higher the reputation level, the more trustful is the information it asserts. The level depends on the credit and reputation that each peer has managed to gather. In Fig. 2, nodes are numbered from A to E, and the color represents the reputation level: from white (*Others*) to black (*ICSIRT*). Specifically, an entity placed in the *Others* level has lower reputation and lower credit value to sign or evaluate an asset committed to the channel by another entity than a entity placed one tier above (e.g., *Sec. Researchers*). Furthermore an entity which earns a certain level of reputation moves up in the tier system, notifying the rest of the network that it is an entity that shares and collects reliable intelligence.

When the system first starts, there is only a single channel, responsible for the coordination between peers. This channel, designated Control channel, is accessible to all the peers and used for membership management, asserting content and checking reputation. The Topic channel corresponds to a specific context, such as "Hardening" or "Vulnerabilities". A peer starts by placing information in a specific topic (1), which executes a transaction in the ledger. This starts the execution of smart contracts in all the remaining peers (2), according to the policies defined in the system. Each smart contract will then check the content and assert its validity (3). When a peer wants to read, it checks the topic (4) and asserts its validity by checking the control channel (5).

Within the network, it is possible to create more channels (Fig. 3). In a
stationary situation, peers are constantly interacting with the control channel
and, eventually, specific topic channels (1). When necessary, a participant can
deploy a new channel by sending a request to the Platform (2). This will then
handle the deployment of the genesis block for the channel intended (3) and
communication can start in the new channel (4).

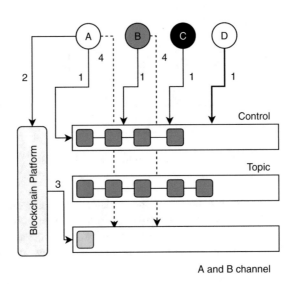

Fig. 3. Deployment of a private channel

These channels can be used in two perspectives. They can be a topic channel,
which then originates an open channel for all the participants who want to share
intelligence, i.e., a channel dedicated to the submission of vulnerabilities. On
the other hand it can be used as a private channel, for confidentiality purposes,
between two or more organizations. The creation and deployment process occurs
in the same way in both cases, although the channel configurations like policies
and chaincode may differ.

It is not imperative for the proper functioning of the system that entities
join in private channels. These private channels are new instances of a ledger,
that can hold new chaincode and policies but, in a more theoretical sense, they
are channels of communication seen only by those who have access to it. They
are a kind of secret rooms inside a open house where only few who want to do
business in the "dark" know its whereabouts.

3.1   Implementation Details

The implementation and validation of the architecture relied on Hyperledger
Fabric, where its unique features, fit perfectly into the main requirements of this

work. Furthermore, being a highly modular platform, allows developers to adapt the solution for their needs. Fabric does not impose the use of tokens or digital currency to work. Moreover is an open source platform, actively developed and with a strong and active community.

Transactions are key in Fabric and are of the responsibility of the ordering service. A transaction starts with a proposal, sent from the client and received by the peer component in the network. The peer then simulates the execution of that transaction against the current version of the values in the state or world state of the ledger (Fig. 4). The result of this simulation is a *read/write* set. Since all peers are synchronized, they should execute the proposal with the same values and obtain the same result. The peer now signs the transaction and sends it back to the client in a endorsement response format along with cryptographic materials.

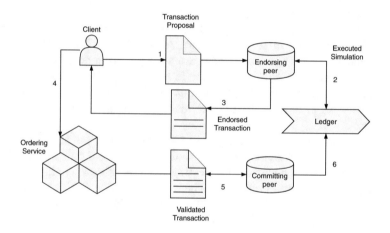

Fig. 4. Fabric consensus in transaction lifecycle

The client sends a invocation request to the ordering service onward with the endorsed response that will be analyzed against the policies defined in the network. If the transaction does not meet the established criteria, it will not be classified as valid and it will not update the state of the ledger, rather, it will be stored in the blockchain part of the ledger. However, if it meets the policies and if all endorsement responses from the peers have the same result, it is assigned as a valid transaction and submitted to the ledger to update the world state.

The endorsement policies are also a key component, by determining if a transaction is valid or not. In the most common endorsement policies it is usually specified that for a transaction to be valid it must be endorsed or signed by a specific set of peers or by the majority of the peers [5]. Only designated administrators from the network have the permission to update or modify this policies that are agreed by all network participants.

4   Test Scenario

The Fabric platform uses docker containers as the baseline for their components deployment. Therefore, it was also necessary to implement docker orchestration tools. To this end, docker-compose was used, mainly for simplicity.

The test scenario included the development of three organizations ("orgA" - CSIRT, "orgB" - Sec. Researcher, and "orgC" - ICSIRT), were each peer was deployed in a different host, simulating three real world entities and an additional node for the ordering service. In addition, three channels where also created: the control channel, where entities reflect their intention to join the network and create their reputation level; one private channel between the organizations "orgA" and "orgB"; one private channel between organizations "orgA" and "orgC".

The chaincode used to create the tier profile for each organization is deployed only in the control channel. Therefore, entities can only execute queries to obtain other member reputation level in that specific channel.

The first test was used to implement a specific channel for the exchange of the PGP public key fingerprint of each organization. The logic behind this choice is that because PGP public keys are widely used in email encryption and signing, and so trust in this information is paramount. Furthermore, in future services additions to the system, the use of the PGP public key fingerprint will be imperative to execute and interact in those topics, as a way for identity verification.

Peers on both private channels could not execute nor see the chaincode for the reputation profile creation, proofing that chaincode can only be called in the specific channel that is deployed. For testing purposes simple chaincode samples were deployed in the private channels. These created two variables to hold integer values and make simple value transactions between the two variables.

The private channels were not visible for the specific organization that did not had permission, e.g., "orgB" could not see nor make any kind of connection to the private channel between "orgA" and "orgC", and vice versa.

Participants for each organization were able to create their tier profile, give credit to one another for the submission of their PGP public key fingerprint, make simple values transactions executing the chaincode in each private channel and finally, they were also able to request to move up in the tier system once the reputation level was enough to do so.

5   Conclusion and Future Work

The work described in this paper describes an architecture for the storage and exchange of cyber-intelligence using a blockchain. For this, Hyperledger Fabric was used, with the Raft consensus model. Each actor in the communication is granted a reputation level, that increases with the amount of credit received from the other actors. Also, each actor is required to validate the information it finds in the channels and, the higher the reputation level, the higher the trust on the information. This gives an indication to the other actors of the overall quality and correctness of the information present in the channels. The system rewards actors that contribute for the network, with access to privileged intelligence.

The channels are implemented as a distributed ledger, based on blockchain. Each actor can be allowed to access a channel and can also be allowed to create new channels, that can either be public or private.

For proof-of-concept, a test scenario with three organizations was implemented. It successfully demonstrated the architecture for the exchange and validation of information. The initial architecture is still considerably complex to assemble, and this is an issue that should be addressed in the future, along with the choice for the format to share the intelligence used in the system. Since there is no world accepted standard this is still a object in study, although all directions point to STIX 2.0.

Another problem lies in encouraging entities to sign information submitted by the other participants. This can become a problem because if information is not being signed the actor's reputation cannot increase, resulting in a tier stagnation.

References

1. Bavisi, S.: Penetration testing. In: Managing Information Security, pp. 177–200 (2013). Elsevier. https://doi.org/10.1016/B978-0-12-416688-2.00007-6. https://linkinghub.elsevier.com/retrieve/pii/B9780124166882000076. Accessed 27 Jan 2020. ISBN: 978-0-12-416688-2
2. Nurse, J.R.C., Creese, S., Goldsmith, M., Lamberts, K.: Guidelines for usable cybersecurity: past and present. In: 2011 Third International Workshop on Cyberspace Safety and Security (CSS), ISSN: null, 2011 September, pp. 21–26. https://doi.org/10.1109/CSS.2011.6058566
3. Marback, A., Do, H., He, K., Kondamarri, S., Xu, D.: A threat model-based approach to security testing. Softw.: Pract. Exp. **43**(2), 241–258 (2013). https://doi.org/10.1002/spe.2111. http://doi.wiley.com/10.1002/spe.2111. Accessed 27 Jan 2020. ISSN: 00380644
4. Vukolić, M.: Rethinking permissioned blockchains. In: Proceedings of the ACM Workshop on Blockchain, Cryptocurrencies and Contracts - BCC 2017, pp. 3–7. ACM Press, Abu Dhabi (2017). https://doi.org/10.1145/3055518.3055526. http://dl.acm.org/citation.cfm?doid=3055518.3055526. Accessed 28 Jan 2020. ISBN: 978-1-4503-4974-1
5. Androulaki, E., Manevich, Y., Muralidharan, S., Murthy, C., Nguyen, B., Sethi, M., Singh, G., Smith, K., Sorniotti, A., Stathakopoulou, C., Vukolić, M., Barger, A., Cocco, S.W., Yellick, J., Bortnikov, V., Cachin, C., Christidis, K., De Caro, A., Enyeart, D., Ferris, C., Laventman, G.: Hyperledger fabric: a distributed operating system for permissioned blockchains. In: Proceedings of the Thirteenth EuroSys Conference on - EuroSys 2018, pp. 1–15. ACM Press, Porto (2018). https://doi.org/10.1145/3190508.3190538. http://dl.acm.org/citation.cfm?doid=3190508.3190538. Accessed 28 Jan 2020. ISBN: 978-1-4503-5584-1
6. Nakamoto, S.: Bitcoin: a peer-to-peer electronic cash system (2008)

7. Mohanta, B.K., Panda, S.S., Jena, D.: An overview of smart contract and use cases in blockchain technology. In: 2018 9th International Conference on Computing, Communication and Networking Technologies (ICCCNT), pp. 1–4, July 2018. IEEE, Bangalore. https://doi.org/10.1109/ICCCNT.2018.8494045. https://ieeexplore.ieee.org/document/8494045/. Accessed 28 Jan 2020. ISBN: 978-1-5386-4430-0

8. Yaga, D., Mell, P., Roby, N., Scarfone, K.: Blockchain technology overview. National Institute of Standards and Technology, Gaithersburg, MD, Technical report NIST IR 8202, October 2018, NIST IR 8202. https://doi.org/10.6028/NIST.IR.8202. https://nvlpubs.nist.gov/nistpubs/ir/2018/NIST.IR.8202.pdf. Accessed 02 Jan 2020

9. Zheng, Z., Xie, S., Dai, H., Chen, X., Wang, H.: An overview of blockchain technology: architecture, consensus, and future trends, pp. 557–564 (2017). https://doi.org/10.1109/BigDataCongress.2017.85

10. Fischer, M.J.: The consensus problem in unreliable distributed systems (a brief survey). In: Karpinski, M. (ed.) Foundations of Computation Theory. Lecture Notes in Computer Science, pp. 127–140. Springer, Berlin (1983). https://doi.org/10.1007/3-540-12689-9_99. ISBN: 978-3-540-38682-7

11. Ongaro, D., Ousterhout, J.: In search of an understandable consensus algorithm, p. 18 (2014)

12. Howard, H., Schwarzkopf, M., Madhavapeddy, A., Crowcroft, J.: Raft reoated: do we have consensus? In: ACM SIGOPS Operating Systems Review, vol. 49, no. 1, pp. 12–21, January 2015. https://doi.org/10.1145/2723872.2723876. http://dl.acm.org/citation.cfm?doid=2723872.2723876. 02 May 2020. ISSN: 01635980

13. Buterin, V.: A next generation smart contract & decentralized application platform. Technical report, p. 36 (2013). https://www.weusecoins.com/assets/pdf/library/Ethereum_white_paper-a_next_generation_smart_contract_and_decentralized_application_platform-vitalik-buterin.pdf

14. Homan, D., Shiel, I., Thorpe, C.: A new network model for cyber threat intelligence sharing using blockchain technology. In: 2019 10th IFIP International Conference on New Technologies, Mobility and Security (NTMS), June 2019, pp. 1–6. https://doi.org/10.1109/NTMS.2019.8763853. ISSN: 2157-4960

Envisioning the Digital Transformation of Financial Documents: A Blockchain-Based Bill of Exchange

Andrea Ponza[1]([✉]), Simone Scannapieco[1], Anna Simone[2], and Claudio Tomazzoli[2]

[1] Real T S.R.L., 37131 Verona, VR, Italy
{andrea.ponza,simone.scannapieco}@realt.it
[2] University of Verona - Department of Computer Science, Strada le Grazie 15, 37134 Verona, VR, Italy
anna.simone@studenti.univr.it, claudio.tomazzoli@univr.it

Abstract. A Bill of Exchange (BoE) is a paper-written contract involving three parties A, B and C where A is economically in debt with B and in credit with C. Once the parties approve a BoE, C is legally bound to pay B on behalf of A within a set deadline, so that the debt of A towards B is extinguished. Although regarded as an elegant and powerful variant of a promissory note, over time the BoE has become unpractical to use in a global market where suppliers and customers aren't next-door companies anymore. On the other hand, the blockchain distributed ledger, AES authentication, and digital archiving with suitable long-period standards (e.g., PDF/A) may encourage the revival of such an instrument, while ensuring legal validity, strength and a non-tampering warranty.

This paper exploits said state-of-the-art technologies to bring the paper-based BoE into the digital era as the DigiBoE. Its envisioned applications are B2B, C2C and B2C secure and legally acknowledged transactions for debt resolution no longer requiring financial intermediaries.

Keywords: Blockchain · Bill of exchange · Smart contracts · FinTech · DeFi

1 Introduction and Motivations

Traditional industry and finance are undergoing a deep metamorphosis. The integration with brand new enabling technologies allows the rebuilding of internal business models boosting productivity and offer quality. Notably, *blockchain* is broadly discussed as a paradigm carrying huge cross-sectoral innovation potential, for which the research community seeks effective applications. Blockchain-based *smart contracts* (e.g., for insurance policies [10,19]) drastically reduce administration costs and increase internal processes' efficiency and automation. Blockchain has also been used to certify the proof of existence and authorship of a

J. Prieto et al. (Eds.): BLOCKCHAIN 2020, AISC 1238, pp. 81–90, 2020.
https://doi.org/10.1007/978-3-030-52535-4_9

document [4,17], thus ensuring intellectual property and non-tampering [16,20].

As a matter of fact, the first successful application of blockchain technology is represented by 2009's *Bitcoin*, which then rapidly affected the entire financial sector. Any economic transaction between people (such as stock market trades [14]) could be digitalized and verified through blockchain without intermediaries. Based on this principle and on the examples above, it seems reasonable to argue that, under the proper conditions, almost all contract types regulating such transactions could be safely represented and stored in a public encrypted ledger.

In the context of this paper, the focus is the *Bill of Exchange* (hereafter, BoE): a popular financial tool of the past. A BoE is an agreement—traditionally pledged on paper—and a variant of the *promissory note*. As businessmen know, a promissory note enforces a legal bond between two parties: a *debtor* emits and signs a document to defer a payment owed to a *creditor* until a set deadline. The creditor accepts the payment deferral with his counter-signature. Clearly, both debtor and creditor benefit from the accord: the former may better organize their finances, while the latter is safeguarded if the debtor evades the contract's terms. Unlike a promissory note, a BoE splits the *drawer*, who emits the contract, from the *drawee*, who is requested to extinguish the debt, keeping the *recipient* as the creditor. When entity A is in debt with entity B and in credit with entity C, a BoE emitted by the drawer A and accepted by C (the drawee) enforces C to pay B on behalf of A. In other words, the debtor shifts from A to C (holding B as the recipient), and A's debt towards B is extinguished.

BoEs shone in the distribution sector, where retailers evaded insolvency with habitual suppliers by designating a payer *other than* banks, money lenders, or other financial intermediaries. However, e-commerce grew over time as the undisputed way to buy assets worldwide, whilst distancing consumers and providers. The standard, paper-written BoE—where legal validity is ensured by handwritten signatures—could not stand the pace, becoming impractical and slowly abandoned due to such a massive technological leap.

This paper explores several enabling technologies that integrate the blockchain public ledger, to possibly bring the paper-based BoE into the digital era as the DigiBoE. The already mentioned smart contracts may stand out as a viable enabler for the DigiBoE semantics. In a world moving towards pandigitalization, the DigiBoE could be an agile and lawful debt clearance means. The DigiBoE could be useful in B2B, C2C and B2C environments, especially for SMEs and start-ups with regular liquidity problems or denied loans, while reducing the need for financial intermediaries and providing an equivalent security level.

The rest of the paper is organized as follows. Section 2 reviews the literature about the BoE and known digitalization attempts. Section 3 analyzes the enabling technologies supporting a successful digitalization of the BoE. Section 4 reports a description of the idea and inner workings of the DigiBoE. Section 5 draws some conclusions and explores future work.

2   Related Work

The majority of BoE literature is from the first half of the 20th century, focusing on its birth [18] and usage through the centuries [3,12]. This reveals the important role of the BoE before the digital era, when the globalized market originated its decline. In [9], the authors deal with the BoE's potential in the digital age, limited by three main issues: (i) the need for writing; (ii) the need for signatures; (iii) the admissibility of electronic BoE as evidence in a court of law. It is argued that the writing constraint is satisfied because an e-bill (DigiBoE) would manifest itself either as a reproduction on a computer screen or as a print out of a computer screen. While endorsing the solution of the first issue, this paper addresses the remaining two in the following sections of this paper.

The only notable attempt of BoE digitalization is by Cryptonomica [2], which implemented a very sophisticated blockchain-based framework. In their case, the concerns pointed out by [9] are solved by centralizing authority for arbitration and signature certification. This is quite in contrast with the blockchain's "decentralization" mainstay, for two main reasons: (i) their smart contract includes an arbitration clause linking to their international arbitration online court, shaping Cryptonomica as the *sole conflict solver* in case of legal disputes, meaning their digital BoE isn't a self-valid proof in a public courthouse. (ii) users need to trust Cryptonomica's *centralized* system for public key certification, lacking a guarantee that the signatory has the signing means under its sole control.

3   Enabling Technologies

This section presents the technologies chosen to support a legally valid DigiBoE. The coming analysis citing the Italian legislation can be extended via the *hierarchy of sources of law*, which states that no inferior grade source of law may contrast a superior one. Italian code is therefore bound to be EU compliant. Moreover, lex superior imposes that all European member states have to be similarly compliant.

Blockchain Technology. Blockchain is a public ledger: anyone can check it but cannot modify or erase it. It records digital events among members over a shared network, and can store anything of (not only financial) value. After its validation, a transaction is added to a (possibly new) block linked chronologically to the previous, in a sequence connecting to the original *Genesis Block*. Erasing or altering a block once another is added is deemed infeasible: the more appended blocks, the harder the upkeep for an "alternative reality"-blockchain. No central authority administers this unique decentralized public register: consensus regulates Internet-based peer-to-peer transactions between two individuals. While similar to a *trustless* mechanism, this is, in a focus shift, minimizing the trust required from transacting parties: the underlying technology is simultaneously ensuring the authenticity of the sender and the validity of the currency.

Decentralization means less privacy, but limiting the number of users that can access the network (and thus their informations) may mitigate the issue.

Pursuing decentralization and immutability made the blockchain vulnerable to scalability and speed issues too. On-chain (1st layer) and off-chain (2nd layer) solutions, distributed ledgers and novel consensus mechanisms exist to solve this dilemma. The Lightning Network [15] is a 2nd layer solution allowing user-generation of interpersonal channels via a single blockchain transaction. The connections between said channels create a secure, new network where faster transactions can happen. State channels are similar to Lightning Network channels: on top of payments, they also support a more general *state update*.

Smart Contracts. A contract is any agreement between parties which regulates obligations among them. A *smart* contract is software stored in the blockchain which executes its terms when some conditions are met.

Ethereum, the second most famous blockchain, is one of many open, decentralized platforms running smart contracts. Unlike many, it uses a blockchain to synchronize and store the system's state changes (it is stateful). Smart contracts are written in high-level languages and compiled to bytecode for execution on the *Ethereum Virtual Machine*. Any high-level language could be used, but the most popular languages are tailor-made for writing smart contracts (e.g. *Solidity*).

Ethereum provides two types of accounts: *Externally Owned Accounts (EOAs)* and *Contract Accounts (CAs)*. CAs have no private key, they are controlled by the logic of their smart contract, whereas EOAs have one, and it allows to manage funds and contracts. Ethereum smart contracts execute only when called by transactions (i.e. signed messages originated by EOAs). Contract-creating-transactions need only to contain the contract's compiled bytecode since the "destination" field is used to distinguish contract-creating- from monetary-transactions: the former have a 0x0 (the *zero address*), whereas the latter have the recipients' address. A contract cannot be modified but it can be deleted, and it can call another contract (but the first in the chain must be called by a transaction from an EOA).

Advanced Electronic Signature (AES). AES was introduced in Italy in art. 1, comma 1, lett. q-bis of D. Lgs. 82/2005 [6], and revised by EU Reg. 910/2014 [8] (*eIDAS - Electronic IDentification, Authentication and Signature*), adopted in D. Lgs. 179/2016 [5]. With *QES (Qualified Electronic Signature)*, AES is one of two viable candidates to guarantee an equivalent legal bond between the paper BoE and the DigiBoE. By requiring the signatory to obtain a qualified digital certificate created via a *Qualified Signature Creation Device*, QES is more of a burden than an advantage: both can be used as legal proof in a courthouse, both are under the sole control of the signatory, but AES is available internationally.

In fact, art. 26 from eIDAS specifies that «an advanced electronic signature shall meet the following requirements: (a) it is uniquely linked to the signatory; (b) it is capable of identifying the signatory; (c) it is created using electronic signature creation data that the signatory can, with a high level of confidence, use under his sole control; (d) it is linked to the data signed therewith in such a way that any subsequent change in the data is detectable». These requirements make AES an authentication method recognized in legal courthouses across the EU. It represents one of three requirements for the validity of DigiBoEs.

Multisignature Address. Most Bitcoin addresses require "single-signature transactions", the one signature of the private key owner. Blockchain technology supports however more complicated transactions, involving more than one party for funds mobilization. For instance, *M-of-N multisignature addresses* are born from the cooperation of people, institutions or software. A 2-of-3 could be a parents' savings account for their child: the child can use the funds with one of the two parents' approval, and money can't be withdrawn from the parents if both don't agree. N blockchain addresses and their public keys are algorithmically combined creating a multisignature address, where M private keys allow immediate use.

PDF/A. The *Portable Document Format (PDF)* was born in 1992 from Adobe to help the revolution from paper to digital documents. In 2005 the ISO committee derived the *PDF/Archiving* (shortened in PDF/A) substandard to safeguard long term document archiving. The format conformity conditions are written and periodically updated from ISO in the 19005 family of standards. The current version is ISO 19005-1:2005(E) [11] described in art 2–6, i.e. PDF/A-1.

Art. 6 of Italian DPCM 03/12/2013 [7] declares three document storage formats: XML, plain text, and PDF/A. The third is a standard de facto for document sharing and storage, and it's already used in courts—unlike the other two. It can be digitally signed and could grant the DigiBoE a readable interface.

4   Blockchain Applied to the Bill of Exchange

The following potential approach confirms the technical feasibility of the project. It is thus expected to obtain an ex ante evaluation of the impact of the envisioned results on the competitiveness of the beneficiary subject, thanks to pinpointing of the minimal functional standards required for the application.

As already pointed out, the BoE is a strong and simple tool, easy to understand but obsolete and impractical to use in a market where customers and suppliers aren't next-door companies anymore: with its digitalization they would have an easier time dealing with their debts. Nevertheless, a crucial requirement for DigiBoEs is to ensure the same legal validity of their paper counterpart: they have to have the same juridical power as an old-style BoE, especially if used as proof of missing payment or debt extinction in a courthouse. Preliminary legal analyses showed that the aforementioned requirement is ensured only if:

- The digital support for the bill of exchange adheres to the PDF/A standard;
- The digital support is signed with an Advanced Electronic Signature (AES);
- Non-tampering of both the document and the signatures is guaranteed.

The advantage of using blockchain technology is clear: it plays a pivotal role as the strongest state-of the-art-technology for a non-tampering warranty. Its ledger makes it possible to jointly create, evolve and keep track of events over a shared network and, being public, it can be checked by anyone and it cannot be modified or erased. As should be clear by now, this project wouldn't be feasible without the blockchain's distributed ledger, which provides an efficient way to record transactions in a verifiable and, most of all, permanent way.

Blockchain technology comes with important features such as the *semantic ledger* and smart contracts: when on the blockchain, new information acquires a much deeper meaning beyond that of simple stored data, and it may become a precondition to trigger the execution of parts of an application on the blockchain. In fact, this solution relies on the robustness of smart contracts, and the multisignature address mechanism: together they shall provide a sufficiently robust technology to create a digital equivalent of the paper-based BoE. If scaling or transaction speed ever becomes an issue, the transaction can be moved to state channels which will create a second layer for chained DigiBoEs to happen on. However, these technicalities will be hidden to the end user, regarded as implementation details that contribute to the application being secure, safe and legally binding, but making it accessible and straightforward to use at the same time.

Figure 1 is explained by layer (Sect. 4.1) and following its flow (Sect. 4.2).

4.1 Architecture

Front-End. Subdivided in two macro-blocks:

- *interface*: a user can fill in and submit an AES-signed DigiBoE (drawer), verify a received DigiBoE and authorize it with an AES signature (drawee) or be acknowledged as beneficiary of a digital bill of exchange (recipient).
- *logic*: supports sensitive data encryption, esp. for CRUD operations (*Create, Read, Update, Delete*) over data submitted to or coming from the back-end. It also offers hash generation to unambiguously identify the PDF/A generated from the multisignature address, the transaction amount, and other metadata.

Core. The encrypted data from a drawer-signed DigiBoE triggers the execution of a smart contract that in turn generates and stores a multisignature address from the public addresses and keys of the involved parties. It simultaneously emits the legal document for verification by the drawee; the drawee's authorization validates the now legally binding document which, together with its hash, triggers an update to the record with the multisignature address, gets encrypted, stored in the database, and an acknowledgment notice is sent to the recipient. The document's hash and the multisignature address are permanently stored on the semantic blockchain as immutable proof of the event.

Back-End. Comprehends all CRUD functionalities with the database and storage (both fully encrypted) in such a way to support the rest of the process.

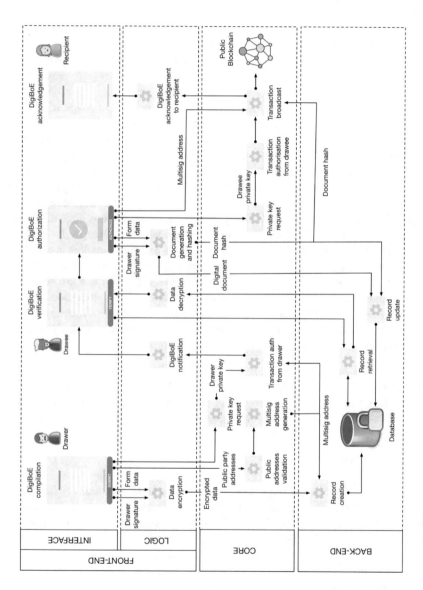

Fig. 1. Flowchart representing the architecture and the process flow for the DigiBoE

4.2   Process Flow

The practical use of a DigiBoE can be thought of as exemplified in this scenario:

1. *Amy* has a € 42 debt towards *Chad* and among her creditors she can count on *Bob*, whose debt towards her is € 42 as well;
2. *Amy* decides to use a digitalized bill of exchange to pay her debt towards *Chad*, trying to get the payment from *Bob* within a set deadline;
3. *Amy* fills in a DigiBoE module with the amount, a deadline proposal, one of her blockchain addresses, *Bob*'s and *Chad*'s email addresses (which will lead to one address for each account), and other metadata;
4. *Amy* submits the DigiBoE to the system with her private key signature: she officially becomes the drawer, proposing *Bob* as drawee and *Chad* as recipient;
5. the three addresses are validated and combined in a multisignature address:
   5.1 *Amy*'s AES and the module's data are encrypted and used to create a DB record in the back-end layer together with the multisignature address;
   5.2 simultaneously, the bill of exchange is sent to *Bob* to get his signature: he verifies it and decides if he wants to sign it—thereby confirming he will pay € 42 within the deadline *Amy* proposed, or proposing himself a date change—or not; if he doesn't accept, the process is interrupted;
6. with *Bob*'s AES, the DigiBoE is binding, and a legal document is generated:
   6.1 the transaction is broadcasted and cemented on the public blockchain;
   6.2 at the same time, this transaction's record is updated in the DB with the document signed by *Amy* and *Bob* and the hash of the document itself;
   6.3 concurrently, *Chad* will now know when his € 42 credit from *Amy* will be settled, thanks to a reception confirmation; however, it isn't strictly necessary to communicate that it is *Bob* who will pay.

Finally, if *Bob* were to have a relevant credit with *Dawn*, he could himself generate a new DigiBoE—a chained DigiBoE—asking her to pay his new debt with *Chad*. Chaining is in fact a prominent characteristic of the old BoEs where state channels could shine, creating a faster throughput in a single transaction.

5   Conclusions

This project contributed, through the use of blockchain technology, to find a way to help the pandigitalization movement of recent global interest. This study has in fact investigated one of the possible functioning processes allowing the digitalization of the BoE. This study's main result is a new and modernized technological counterpart for the BoE. Propagation of the BoE as feasible model for the 21st century marketplace can be achieved through blockchain technology.

Three are the requirements found to be the minimum for digitalization:

- a long-term digital support like the PDF/A standard;
- an internationally valid signature like AES;
- a non-tampering guarantee like the stateful, smart-contract enabled Ethereum.

This is by no means the only viable triplet required for an implementation, which is itself left for future work. As explored in Sect. 3, plain text and XML can substitute PDF/As, and QES could replace AES. As Cryptonomica demonstrated, foregoing decentralization is also an option, by imposing a central authority for signature validation (and arbitration). Ethereum itself is not the only viable blockchain implementation that allows for smart contracts and state storage: Æternity [1] is a more scalable alternative with state channel support, and NEO [13] is a lightweight alternative which is lacking widespread adoption.

Being a web-based distributed technology, the expectation is for the users to promote DigiBoE adoption, driven by how it solves frequently encountered problems and the value it gives. The main customers are active and productive SMEs which have trouble to get credit from banks. It might be the limited amount of credit requests that bank allow or the hard and complex inquest process faced in particular by companies, but the restricted credit opportunities will be greatly enhanced, and thus advertised to more potential users in the process.

References

1. Æternity (2017) æternity - a blockchain for scalable, secure and decentralized æpps. https://aeternity.com/
2. Ageyev, V., Baryshnikov, M., Kurylovych, V.: Cryptonomica white paper (2018). https://github.com/Cryptonomica/cryptonomica/
3. Ashton, T.S.: The bill of exchange and private banks in Lancashire, 1790–1830. Econ. History Rev. **15**(1/2), 25–35 (1945). https://doi.org/10.2307/2590309
4. Bodó, B., Gervais, D., Quintais, J.P.: Blockchain and smart contracts: the missing link in copyright licensing? IJLIT **26**(4), 311–336 (2018). https://doi.org/10.1093/ijlit/eay014
5. D. Lgs. 179/2016: Decreto Legislativo 26 agosto 2016, n.179. Repubblica Italiana, Modifiche ed integrazioni al Codice dell'Amministrazione Digitale, di cui al decreto legislativo 7 marzo 2005, n.82, ai sensi dell'articolo 1 della legge 7 agosto 2015, n.124, in materia di riorganizzazione delle amministrazioni pubbliche (2016)
6. D. Lgs. 82/2005: Decreto Legislativo 7 marzo 2005, n.82. Repubblica Italiana, Codice dell'Amministrazione Digitale (2005)
7. DPCM 03/12/2013: Decreto del Presidente del Consiglio dei Ministri 3 dicembre 2013. Repubblica Italiana, regole tecniche in materia di sistema di conservazione ai sensi degli articoli 20, commi 3 e 5-bis, 23-ter, comma 4, 43, commi 1 e 3, 44, 44-bis e 71, comma 1, del Codice dell'amministrazione digitale di cui al decreto legislativo n.82 del 2005 (2013)
8. EU Reg. 910/2014: European Regulation July 23rd, 2014, n. 910. European Union, regulation (EU) No 910/2014 of the European Parliament and of the Council of 23 July 2014 on electronic identification and trust services for electronic transactions in the internal market and repealing Directive 1999/93/EC (2014)
9. Gamertsfelder, L.: Electronic bills of exchange: will the current law recognise them? UNSWLJ **21**(2), 566–577 (1998)
10. Gatteschi, V., Lamberti, F., Demartini, C., Pranteda, C., Santamaría, V.: Blockchain and smart contracts for insurance: is the technology mature enough? Fut. Internet **10**(2), 20 (2018). https://doi.org/10.3390/fi10020020

11. ISO 19005-1:2005(E): Document management – Electronic document file format for long-term preservation – Part 1: Use of PDF 1.4 (PDF/A-1). Standard, International Organization for Standardization, Geneva, CH (2005)
12. Maixé-Altés, J.C., Iglesias, E.M.: Domestic monetary transfers and the inland bill of exchange markets in spain (1775–1885). JIMF **28**(3), 496–521 (2009). https://doi.org/10.1016/j.jimonfin.2008.11.001
13. NEO (2014) Neo smart economy. https://neo.org/
14. Po, C., Po, C., Marce, A., Ves, A., Petrica, T., Cioar, T., Anghe, I., Salomi, I.: Decentralizing the stock exchange using blockchain an ethereum-based implementation of the bucharest stock exchange. In: ICCP 2018, IEEE, IEEE, Cluj-Napoca, Romania, pp 459–466 (2018). https://doi.org/10.1109/ICCP.2018.8516610
15. Poon, J., Dryja, T.: The Bitcoin Lightning Network: Scalable Off-Chain Instant Payments, DRAFT Version 0.5.9.2 (2016)
16. de la Rosa, J.L., Gibovic, D., Torres, V., Maicher, L., Miralles, F., El-Fakdi, A., Bikfalvi, A.: On intellectual property in online open innovation for SME by means of blockchain and smart contracts. In: WOIC2016, Barcelona, Spain (2016). https://doi.org/10.13140/RG.2.2.27099.57124
17. Savelyev, A.: Copyright in the blockchain era: promises and challenges. CLSR **34**(3), 550–561 (2018). https://doi.org/10.1016/j.clsr.2017.11.008
18. Usher, A.P.: The origin of the bill of exchange. JPE **22**(6), 566–576 (1914). https://doi.org/10.1086/252472
19. Vo, H.T., Mehedy, L., Mohania, M., Abebe, E.: Blockchain-based data management and analytics for micro-insurance applications. In: CIKM 2017, pp. 2539–2542. ACM (2017). https://doi.org/10.1145/3132847.3133172
20. Wang, J., Wang, S., Guo, J., Du, Y., Cheng, S., Li, X.: A summary of research on blockchain in the field of intellectual property. In: IIKI 2018, vol. 147, pp. 191–197 (2019). https://doi.org/10.1016/j.procs.2019.01.220

Data Protection Compliance Challenges for Self-sovereign Identity

Alexandra Giannopoulou$^{(\boxtimes)}$

Institute for Information Law (IViR), University of Amsterdam,
Amsterdam, The Netherlands
a.giannopoulou@uva.nl

Abstract. Various identity management solutions are emerging in different jurisdictions, with the goal of creating a unified and privacy-preserving identity management system bridging the offline with the online. Within this trend, the concept of self-sovereign identity has re-emerged. It is a concept attached to expressions of both individual autonomy and individual control (sovereignty)—an aspiration in direct relation to what blockchain is promised to bring in contemporary discourse. The paper will provide an overview of the current self-sovereign identity paradigm solutions within the technological environment that involves decentralized networks and it will trace some of the challenges it faces within the European Union especially with regards to the General Data Protection Regulation (EU) 2016/679 (GDPR).

Keywords: Self-sovereign identity · GDPR · Data protection

1 Introduction

Humans are just the sort of organisms that interpret and modify their agency through their conception of themselves.

This is a complicated biological fact about us.
Amelie Rorty

The European Commission's new data strategy has qualified data as 'the lifeblood of economic development'. Naturally, it highlights the innovation potential and the 'transformative' prospects of facilitating data flows while also pointing out the importance of maintaining the right to privacy. However, at the same time, achieving individual "data empowerment" or "data sovereignty" is becoming progressively more challenging without the appropriate technological tools. Since their popularisation, distributed ledgers have reinforced the push towards alternative data governance through decentralized technologies that require a different logic of intermediary entities. Described as foundational technology, the blockchain created the ideal environment for new data ordering regimes to flourish; a disruptive technology with innovation potential for data sharing, money influx, and institutional curiosity.

J. Prieto et al. (Eds.): BLOCKCHAIN 2020, AISC 1238, pp. 91–100, 2020.
https://doi.org/10.1007/978-3-030-52535-4_10

Currently, there is no shortage of blockchain projects that promise to revolutionise both the way that data circulates and individual negotiating power within that network. Among these projects, 'self-sovereign identity' ones have been steadily gaining momentum.

Identity management systems, notably those built in the form of decentralized ledger applications, have been on the forefront of the technological innovation agendas of public institutions, private companies, and privacy-aware communities. This contribution will provide an overview of the current self-sovereign identity technological paradigm solutions and the challenges they face within the European Union especially with regard to the General Data Protection Regulation (EU) 2016/679 (GDPR).

Overall, there is no uniform rule about what constitutes a person's identity, since the concept and its governing norms shift depending on the legal, technological or institutional context. In the past decades, the digital expansion of ourselves has shaped the idea of the creation of a "digital identity". Moreover, the overproduction of personal data in the current data-intensive technological environment that has formed our data-driven societies has created a newly found interest in preserving privacy and data protection online. For example, cases such as that of Cambridge Analytica have illustrated that there are significant shortcomings in the current data management and data governance practices.

The creation of a new -and uniform- digital identity ecosystem is an aspiration that has progressively risen into prominence in various ways and from multiple entities and stakeholders. As a result, various identity management solutions are emerging in different jurisdictions, with the goal of creating a unified, privacy-preserving identity bridging the offline with the online. The market of digital identity is already quite substantial and very diverse. It aims to provide a technological solution to financial inclusion, reputation management, privacy-preserving social media identities etc. Within this trend, the concept of self-sovereign identity has re-emerged. No consistent definition of the concept has been established. In general, we can describe self-sovereign identity as an identity management system. This system developed by a private or a public entity which takes technological design decisions guided by a set of principles that are loosely defined and not universally accepted as a common standard. Overall, it is essentially a technological solution which transcribes the goal of autonomy and individual control of individuals' digital data through decentralization and "user-centric design".

This concept expresses both individual control (sovereignty) and trusted verifiability; familiar aspirations, which blockchain promised to bring to contemporary data protection discourse. In fact, many identity management solutions currently in production rely on the use of decentralized ledgers, cryptography, and local processing of data. These technological design options aim to actualize some of the core principles of lack of central authority to control the identity data, of decentralized verifiability and privacy [1]. However, existing self-sovereign identity implementations choose to prioritize only certain aspects of the intended design principles. In general, there has been a progressive distancing of current technological state of the art from the self-sovereign ideological underpinnings. This has resulted in the interchangeable use of the terms 'self-sovereign identity' and 'decentralized identity' both of which still lack commonly agreed definitions.

With coinciding objectives and features, the development of self-sovereign identity projects has become inseparable from blockchain technological advances and mainstream adoption efforts. However, blockchain proponents have a vested interest in the success of self-sovereign identity solutions as it would constitute the first widely used implementation of blockchain technology.

The eIDAS Regulation defines levels of trust services and thus provides the regulatory environment that enables the creation of different legally compliant identity systems solutions. In addition, identity providers have to conform to data protection regulation such as the GDPR. Compliance with these aforementioned European regulatory frameworks appears to be challenging. This is due to the foundation of modern identity proposals on decentralized technologies, which would add a factor of complexity. Moreover, the privacy by design normative obligation could be potentially ensured but not without a level of uncertainty of the compliance of the chosen technological tools and architectures.

Another important aspect of legal compliance lies on the fact that many applicable legal norms are domain-dependent, with certain areas being highly regulated (i.e. financial markets and institutions). The coexistence of these diverse legal obligations appears to provoke tensions between the applicable legal rules and the objectives pursued by the technology; these tensions would be difficult to reconcile.

The paper will focus on the compliance challenges that self-sovereign identity solutions face when operating within the scope of territorial application of the GDPR and the obligations that this entails. On the one hand, we discuss the shared vision that blockchain technologies and self-sovereign identity have and on the other, we present some of the key compliance mechanisms that self-sovereign identity solutions will need to address in order to offer a product that delivers on its promises. We conclude with suggestions on conciliations between the data protection regulatory framework and then GDPR.

2   The Shared Vision of Control

Visions of a self-sovereign self-have been attached to different political ideologies. It was only in 2016 that the fundamental design principles of the concept of self-sovereign identity came to life in a form of a check list of design options by cryptographer Christopher Allen. In his blog[1], he describes the core principles of an identity ecosystem controlled by each individual, which does not hinge on a specific powerful technological infrastructure nor a private or public entity. After tracing the evolution of identity management systems -from centralized to federated to user-centric- the author points out that the self-sovereign identity goes a step beyond the previous systems in that it prioritizes user autonomy through ten foundational principles: existence, control, access, transparency, persistence, portability, interoperability, consent, minimalization, and protection. These principles ensure that the user remains the sole gatekeeper of

---

[1] Allen, C. (2016), The Path to Self-Sovereign Identity, 25 April 2016, Life With Alacrity blog, http://www.lifewithalacrity.com/2016/04/the-path-to-self-soverereign-identity.html.

their respective personal data that constitute the identity that actors will seek to user in order to provide respective services. The concept is goal oriented, focused on preserving 'the right for the selective disclosure of different aspects of one's identity and the various components thereof, in different domains and contextual settings' [4].

From a legal standpoint, the General Data Protection Regulation[2] was designed to provide the legal framework with the appropriate assurances that enable individual control over personal data. As a matter of fact, recital 7 of the GDPR directly highlights that "natural persons should have control of their own personal data". This control principle is conveyed through a set of accountability measures that are imposed on responsible actors and a set of rights assigned to each data subject in order to empower them to exercise control over their personal data. The technological solution of the decentralized identity relies heavily on informed consent, as the regulatory representation of the expression of the autonomy of natural persons [5].

From a technological standpoint, an essential aspect of decentralization is that "no trusted third party should be given control of data, but instead individuals and groups should maintain control over their own data" [2]. *A priori*, distributed ledger technology is aiming to embody the principles of control, security, and transparency. Currently, one of its most prominent promises remains 'data sovereignty'. Sullivan highlights that blockchains could enable individuals 'to control access to their identity information and to create, manage and use a self-sovereign identity' [3]. Consequently, blockchains provide the technological guarantees for trusted data sharing because they permit modularity, transparency and security though encryption. Lately, the techno-legal and ideological circumstance has guided the flourishing of a market that proposes diverse identity solutions on a decentralized environment. Depending on how they are designed, these solutions can convey a granular level of decentralization [6]; thus, not all decentralized identities are similar. Finally, the variety of the solutions is due not only to the diverse technological design choices but also to the entity that develops them.

With the principles of a decentralized identity being constantly in flux, the standardization of technological architectures constituted a first effort towards harmonization. Among the existing solutions, the W3C has launched a set of standard setting processes for decentralized identity in order to provide a unified strategy towards a common aspiration of eradicating centralized control of personal data. Still, whether the consortium will succeed in enabling a decentralized identity infrastructure or whether the mistakes of previous standardizing attempts on the Web will be repeated remains unclear. 'Will the blockchain revolution bring a new decentralized web into existence, or simply become the technical infrastructure of further control and centralization?' [2].

According to the W3C existing decentralized identity technical documentation [4], 'decentralized identifiers (DIDs) are a new type of identifier, and they are an essential component of decentralized digital identity. Furthermore, DIDs are designed to enable

---

[2] Hereinafter GDPR.

the controller of a DID[3] to prove control over it and to be applied independently of any centralized registry, identity provider, or certificate authority'. Also, the technical specifications describe the control of public and private keys by the individual (or data subject in legal terms) through the use of a DID document in order to autonomously manage the information related to them.

The DID document is defined as 'a set of data describing the DID subject, including mechanisms, such as public keys and pseudonymous biometrics, that the DID subject can use to authenticate itself and prove their association with the DID. A DID document might also contain other attributes or claims describing the subject'. The qualification of this data as personal depends on the data protection test of identifiability which ultimately leads to the qualification of data as anonymous or pseudonymous and thus, personal. Seeing as in the context of the semantic Web digital identifiers can be attributed to a multiplicity of entities such as Internet of Things, companies etc., the GDPR compliance questions refer only to those DIDs that are used to manage data that refer to natural persons.

The DID technical descriptions represent one example of decentralized control. Finally, the implementation of these identifiers within the overall identity system will determine the GDPR applicability, actor accountability, and the data subjects' rights enforcement.

3 Rights and Obligations in a Self-sovereign Ecosystem

There is no universal standard for the use of decentralized identifiers. For example, they can be stored in the personal device of the user. Similarly, the verifiable credentials referring to that individual can be put on the distributed ledger in various privacy-preserving forms, and they can also be transmitted to third parties and entities. Overall, maintaining the ledger of transactions of verifiable credentials necessitates a network of nodes that process and collectively keep the distributed database up to date. The pressing compliance questions in terms of the GDPR are two-fold: first, can data published in the -public or private- typically permissioned networks be qualified as personal and second, what are the accountability obligations among the p2p network of nodes. While these questions have been addressed beforehand for public blockchains in general [1], the specific technological set developed for decentralized identities provides a new breeding ground on which GDPR compliance can be examined.

3.1 Personal Data Processing

The generation of the decentralized identity starts with the issuance of a verifiable credential stored at the individual's device, which contains the public and private keys belonging to the user. The keys in question constitute part of the identity of the user and are considered pseudonymous data according to the WP29's opinion, which underlines

---

[3] Which should not be confused with the legal concept of the data controller. It will mostly refer to the data subject, or the owner of the identity. However, in some cases, it can be a third party acting on behalf of the data subject.

that asymmetric encryption methods are pseudonymisation methods that 'merely reduce the linkability of a dataset with the original identity of a data subject, and is accordingly a useful security measure'.

Pseudonymous data are protected by the GDPR according to article 4(5) GDPR, as personal data. According to Recital 30 of the GDPR, 'natural persons may be associated with online identifiers provided by their devices, applications, tools and protocols, such as internet protocol addresses (...) or other identifiers (...) This may leave traces which, in particular when combined with unique identifiers and other information received by the servers, may be used to create profiles of the natural persons and identify them'.

Public keys fall into this category, and their management falls into the risk assessment obligation of data controllers, per the Article 25 GDPR privacy by design obligations. However, the French Data Protection Authority (CNIL) has issued an official opinion on blockchains arguing that 'the very architecture of blockchains means that these identifiers are always visible, as they are essential for its proper functioning. The CNIL therefore considers that this data cannot be further minimised and that their retention periods are, by essence, in line with the blockchain's duration of existence'. Therefore, it can be argued that the public keys, combined with necessary privacy enhancing mechanisms (PETs) could potentially fulfil the data minimisation requirements of the GDPR.

Moving beyond public and private keys, the generation of hashes that serve as attestations of the transaction of verifiable credentials on-chain merits our attention from a data protection perspective. In order to respond to the question of whether these hashes fall under the personal data qualification of the GDPR, one has to refer to the notion of risk is pervasive across the Regulation [8]. For example, in article 25 GDPR, the privacy by design obligation is measured through the concept of risk. As a matter of fact, the data controller(s) obligations to implement technical and organizational measures has to be considered according to the case-specific processing that will take place and in order to minimize the risks for the data subjects' rights. In fact, the goal of the data protection legislation is not to exclude risk or to ensure that it does not manifest in any form during the data processing. Rather, the legislator embraces the risks involved in personal data processing and employs a wide variety of tools (accountability and obligations to responsible actors, data subjects' rights etc.) to minimize the risk involved.

The Regulation encourages pseudonymisation in order to 'reduce the risks to the data subjects concerned'. According to the GDPR, pseudonymisation of data does not equal anonymization. Pseudonymous data are subject to GDPR restrictions. However, the distinction between the two methods is not always clear. The criteria for distinction can be found first on recital 26 GDPR, which specifies that data becomes anonymous if it is 'reasonably likely' that no identification of a natural person can be derived. An individual is considered to be 'identifiable' where they can be 'distinguished' from others. In previous reports, the Article 29 Working Party (which is now the European Data Protection Board) have provided a more absolute interpretation when it comes to various methods of processing of personal data. The two approaches, relative and absolute conflict regularly. The risk-based approach implies that the determination of the risk inherent in the likelihood to re-identify falls on the data controllers.

There is tension between the risk-based factor introduced through the GDPR and the absolute approach that existed thus far. The national DPAs' opinions reflect this lack of homogeneity. For example, according to the Irish DPA, the data have to be rendered 'irreversibly' anonymous, but the criterion of irreversibility is applied in a more relative manner linking it to the absence of reasonable likelihood of identifiability. Similarly, the French Data Protection Authority (CNIL) acknowledges that anonymization tends to make identifiability 'practically impossible'.

Within this normative framework, the A29WP has published an absolute opinion when it comes to hashing as a method of pseudonymization. A more recent report published by the Spanish Data Protection Authority nuances the absolute approach by introducing the notion of risk in its assessment of the technological method. Hence, and according to the Spanish DPA, hashing can at times be considered as anonymization or pseudonymization depending on a variety of factors varying from the entities involved to the type of the data at hand.

According to Recital 26 GDPR, the distinction criteria between pseudonymization and anonymization can be found in the 'the means reasonably likely to be used (..) either by the controller or by another person to identify the natural person directly or indirectly. To ascertain whether means are reasonably likely to be used to identify the natural person, account should be taken of all objective factors, such as the costs of and the amount of time required for identification, taking into consideration the available technology at the time of the processing and technological developments'. The time frame of the appreciation of these factors for the identifiability test is relative according to the GDPR, in conjunction with the available technology.

The risk assessment in the case of recordation of data on blockchains should take into consideration the envisioned timeframe for the technology at hand Within that frame, taking into consideration organisational measures to employ can be a risk-minimizing factor for the obligations that the data controllers are facing. When the issuer presents the hash of the credential on the blockchain, it is important to assess the likelihood of identification according to the person or entity that would try to identify. This assessment will have to include perspectives of third parties and of the data controllers, and possible de-identification brute forcing methods such as content-based reidentification.

3.2 Actor Accountability

According to this Regulation, there are two types of actors whose key role in data processing and whose relationship to the data within the data processing environment leads the European legislator to attribute them a set of obligations and responsibilities. Thus, these liable actors are subject to data protection rules.

According to article 24(1) GDPR, data controllers are responsible to make sure that the data processing remains lawful and in compliance with the Regulation. Additionally, the data controllers have to take all necessary measures so that data subjects are sufficiently informed and have the ability to exercise their data protection rights. This set of obligations is directed at the entities responsible to take 'technical and organizational measures' enforcing the GDPR rules. In particular, article 25 GDPR highlights that *'the controller shall, both at the time of the determination of the means for processing and at the time of the processing itself, implement appropriate technical and organisational measures, such as pseudonymisation, which are designed to implement data-protection principles, such as data minimisation, in an effective manner and to integrate the necessary safeguards into the processing in order to meet the requirements of this Regulation and protect the rights of data subjects'*. These obligations place a high level of responsibility for every actor qualified *de facto* as a data controller. Naturally, data controllers are be liable to pay damages for lack of compliance or when these measures are proven to assume too high a risk towards personal data processing in case of unlawful processing.

Data controller is defined in article 4(7) GDPR as "*the natural or legal person, public authority, agency or other body which, alone or jointly with others, determines the purposes and means of the processing of personal data; where the purposes and means of such processing are determined by Union or Member State law, the controller or the specific criteria for its nomination may be provided for by Union or Member State law*". Strictly as defined by law, the concept of the data controller requires further clarifications in order to fit a decentralized model of data processing. As a matter of fact, the centralized linear practices that depend on clearly defined boundaries between data subjects, data controllers, and data processors are constantly challenged and contradicted in today's technological realities. There is a clear need for interpretations, guidelines, and rules that are progressively created by the regulatory framework and its respective case law, as well as the European Data Protection authorities.

A *de facto* analysis of the circumstances of the data processing will determine the entities identified as data controllers, *regardless of what is stated in a previous written contractual agreement between the different participating entities and actors*. The current scope of application for actors considered to be defining the 'means and purposes of the processing' remains rather broad in order to ensure the applicability of the data protection responsibility allocation in a networked environment.

Especially with regard to joint controllership, the concept remains quite broad. For example, the opinion of the Article 29 Working Party[4] on data controllers has highlighted that the role of the 'controller is to determine who shall be responsible for compliance with data protection rules and how data subjects can exercise the rights in practice. In other words: to allocate responsibility'. Based on the normative responsibility allocation among data controllers and consistent case law since the establishment of Directive 95/46/EC[5], controllers' responsibility cannot be contractually waived by any of the responsible actors. However, in case of joint controllership, data controllers can contractually assign partial responsibility based on distinct stages of data processing. Thus, different degrees of responsibility can be assigned proportionately to the participation of the respective data controller to the data processing. Namely, according to article 26 GDPR, joint controllers have to enter into an 'arrangement'. Nevertheless,

---

[4] Article 29 Data Protection Working Party, 'Opinion 1/2010 on the Concepts of "Controller" and "Processor"' (2010), 4.
   Currently, the responsible body is the European Data Protection Supervisor - EDPS.
[5] This Directive was replaced by the GDPR on 25 May 2018.

and regardless of this 'arrangement', data subjects should maintain the ability to exercise their rights against every joint controller. The same can be derived from most recent and previously established case law from the CJEU. See for example, case C-40/17 (Fashion ID GmbH & Co. KG v Verbraucherzentrale NRW eV), which ruled that joint controllership can be established for specific phases of the data processing and that controllership can be then reallocated to only one of the actors for subsequent stages of data processing.

Against this legal background, the participating nodes within a decentralized identity infrastructure could qualify as joint data controllers for the transactional data that they to verify, store, and put on/off chain. Even if the means and purposes of the data processing and the architectural design rules that will govern the safe and secure data processing are decided in a less decentralized manner by a single entity, the participation on the network can lead to such a qualification especially given the progressively expansive case law on the responsibility allocation of data controllers. However, each joint controller can only be considered responsible within the limits of the data processing they are facilitating. The more pragmatic approach -adopted by the responsible bodies and case law- in the determination of liable actors and the allocation of liability between them signifies that actors can be qualified as joints controllers when they exert 'a decisive influence over the collection and transmission' of the personal data, without necessarily having access to the data in question and where there is joint determination of the purposes of the processing.

The architectures of decentralized identity could also lead to the qualification of the data subjects as data controllers with regard to their own data [9], which is a legal concept that has not yet been tested in court but which has been indirectly suggested through developed case law.

4   Conclusion

Decentralized -or self-sovereign- identity is an emerging concept that should be regarded critically for its purported benefits in providing solutions for issues like private and secure exchange of personal data among actors that do not necessarily trust each other and without the mediation of an institution acting as the certifying authority. Whether it consists of a bottom-up approach to establish community-driven norms and solutions to the systemic problem of data-intensive technologies, or company invest-ments in developing a product that corresponds to similar societal needs, or even public institutions aiming to provide innovative solutions for its citizens, compliance with data protection norms is key.

Legal compliance can be the gateway for a lot of these projects to reach some level of recognition and usability but also it can be the tool that ensures that these projects deliver on their promise to redesign personal data exchange.

Finally, future solutions will stem from both the normative level and the techno-logical level. Interpretations and official opinions from data protection authorities on the technological tools and methods applied in decentralized identities are key for providing legal certainty. The techno-legal cooperation of experts from the design stage of these projects would become substantial in risk assessment and management.

Acknowledgments. The Lab has received funding from the European Research Council (ERC) under the European Union's Horizon 2020 research and innovation programme under grant agreement No 759681.

References

1. Finck, M.: Blockchain Regulation and Governance in Europe. Cambridge University Press, Cambridge (2020)
2. Halpin, H.: Decentralizing the social web. Can blockchains solve ten years of standardization failure of the social web? In: Bodrunova, S.S., et al. (eds.) INSCI 2018 Workshops. LNCS, vol. 11551. Springer, Cham (2019)
3. Sullivan, C., Burger, E.: E-residency and blockchain. Comput. Law Secur. Rev. **33**(4), 460–475 (2017)
4. Wang, F., De Filippi, P.: Self-sovereign identity in a globalized world: credentials-based identity systems as a driver for economic inclusion. Front. Blockchain **2**, 28 (2020). https://doi.org/10.3389/fbloc.2019.00028
5. Giannopoulou, A.: Algorithmic systems: the consent is in the details? Internet Policy Rev. **9** (1) (2020)
6. Bodo, B., Giannopoulou, A.: The logics of technology decentralization: the case of distributed ledger technologies. In: Ragnedda, M., Destefanis, G. (eds.) Blockchain and Web Social, Economic, and Web 3.0: Technological Challenges. Routledge, New York (2020)
7. Gooddell, G., Tomaso, A.: A decentralized digital identity architecture. Front. Blockchain **10** (2020). https://doi.org/10.3389/fbloc.2019.00017
8. Finck, M., Pallas, F.: They who must not be identified - distinguishing personal from non-personal data under the GDPR. International Data Privacy Law, 2020; Max Planck Institute for Innovation & Competition Research Paper No. 19-14 (2020)
9. Edwards, L., Finck, M., Veale, M., Zingales, N.: Data subjects as data controllers: a fashion (able) concept? Internet Policy Rev. (2019)

Revisiting Blockhain Use in Notary Services: An European Perspective

António Pinto[1,2(✉)] and Jorge Silva[3]

[1] CIICESI, ESTG, Politécnico do Porto,Porto, Portugal
[2] CRACS & INESC TEC, Porto, Portugal
apinto@inesctec.pt
[3] Ordem dos Notários, Lisbon, Portugal
jorgebatistasilva@notarios.pt

Abstract. Notary services have long been identified as a recurrent example for dematerialisation through blockchain adoption, but has failed to become a world wide reality. The key issue being the distinct legal frameworks throughout the world. Europe in this context has a more restrictive legal context with regard to blockchain use. In this work, we briefly discuss the European role of the Notary, review the existing European solutions and identify related open issues that are not resolved in the existing solutions.

Keywords: Survey · Blockchain · Europe · Notary services

1 Introduction

Soon after the publication of the paper proposing Bitcoin [14], the usage of the blockchain as a support for immutable transaction recording was noticed by academia, research and business. The blockchain technology is stated to already be in its third version [19]. This later version of Blockchain aims at the applicability of this technology to other areas such as government, health, legal, education, science or others, not limiting its usage to cryptocurrency, finance and goods exchange.

A common blockchain classification is regarding the capabilities of each peer of the distributed network. To this regard, in [22], the authors characterise blockchain as permissionless or permissioned blockchains. Permissionless blockchains are open and decentralised, such as Bitcoin or Ethereum. Any peer can join or leave at any time and can read or write data to the chain. Permissioned blockchain are more controlled and limit the set of peers capable of performing read or write operations.

The broad applicability of blockchain is defended by some authors, but argued by others. Atzori, in [1], addressed the applications of blockchain from

A. Pinto—This work is partially financed by National Funds through the Portuguese funding agency, FCT - Fundação para a Ciência e a Tecnologia, within project UIDB/50014/2020.

J. Prieto et al. (Eds.): BLOCKCHAIN 2020, AISC 1238, pp. 101–110, 2020.
https://doi.org/10.1007/978-3-030-52535-4_11

a political perspective, focusing on the decentralised governance with the help of blockchain. They conclude that its usage is possible and desirable if properly managed and through the use of a permissioned blockchain. More recently, in [22], the authors discuss this applicability to the point of producing a flowchart to help determine if blockchain is the appropriate technology to be used. Similar work was presented in [12] where the authors try to assess if blockchain can be used as a substitute of trusted third parties. They identify criteria that could help decide upon the usage of blockchain or of a trusted third party. They conclude that the use of blockchain as a better substitute of a trusted third party is limited and applicable only to a reduced set of use cases.

One similarity can be found in the previously enumerated works pertaining the blockchain adequacy. While there is significant enthusiasm regarding blockchain adequacy, the reality is that there is no perfect match and the adoption of blockchain must be evaluated carefully on a case-by-case basis.

The notary services is one of the non-financial services identified as an use case application suited for blockchain or, more generally, distributed ledger technologies [5]. In a naive view, a Public Notary must certify who created a document, that the document exists and that is has not been tampered with. These services appear as a natural match to the opportunities and functionalities provided by blockchain technology, because the required authenticity verification can be done with blockchain [3]. The majority of the existing blockchain-based solutions conform to this naive view of the Notary. In contrast, a Notary in Europe has much more complex role in society and, due to this, is still far from being digital.

2   Notary Services

The digitalization of Notary services using blockchain bridges multiples realities such as political, technological and legal. DuPont et al., in [8], addressed the relation between law and blockchain technologies, referring even that two key legal devices used in modern times are related to blockchain. These devices being the ledger and the contract. They evoke the marriage contract as an example. Interestingly is their stating of the fact that legal contracts are a form of assurance by allowing businessmen to manage uncertainty. Moreover, they state that one shortcoming of the new smart contracts is its inability to be broken and contacts are made to be broken.

Notaries legal capacities vary accordingly to jurisdiction, meaning that, depending on the legal framework of each particular country or world region, the types of legal acts permitted to notaries vary. The function of a notary in countries where the legal system is characterised as Common Law (USA) is reduced to a mere identity verification action and usually does not required any legal training such as a degree in law or other any type of degree. In contrast, in countries whose legal systems are designated Civil Law (European Union), notaries are required to have a university degree in Law, plus specialised training, and Governments delegate public powers on them [16].

In terms of implications towards the digitalisation of the notary services, these will have an impact on the portfolio of legal acts offered by a digital notary service. It will either restrict the number and type of acts performed, such as proof of existence in Common Law regions, or opens the possibility to digitalise a larger spectrum of legal acts. If one considers Civil Law regions, such as the majority of European countries, due to greater number of offered notary services that are still not digitalised, there is still space for digitalisation.

The powers conferred to Civil Law notaries enable contracts signed before them to have a presumption of authenticity and legality, varying according to country, and may cover areas as diverse as contracts for the establishment of companies, marriage contracts, sale contracts and wills. These contracts have a legal force which can only be challenged through a judicial court and possess a higher value than other contracts not made before notaries, designated by private contracts. Moreover, the function of a Civil Law notary implies, unlike a Common Law notary, additional verification steps besides the verification of the identify of the parts. The long term archival of the documents, attestation of the will of the parts in performing the contract, the compliance with tax obligations, the urban legality, money laundering control, and the suitability for the transmission of the property (verification if the seller is effectively the holder of the property that will be disposed of), are examples of such.

Current practice for document archival is based on asymmetric cryptography techniques and document signing with specific professional certificates. Lemieux, in [10], assessed the use of blockchain and distributed ledgers as a trusted record keeping system. She states that it is possible to identify threats to the long term archival of documents to be used in legal contexts. Moreover, she states that if this threats are not addressed, it will hamper the adoption of blockchain-based record keeping solutions.

An European trend to include legal support for solutions that make use of blockchain are starting to appear. Italy's government has passed Law No. 12 of 11 February 2019 that converts Law Decree No. 135 of 14 December 2018, also called Simplifications Decree [4]. This law aims to accelerate the use of blockchain and smart contract technology in the legal world by recognising this technologies giving it legal value. Furthermore, by recognising blockchain as a form of electronic time validation, it will serve as a qualified electronic time validation in all other Member States of the European Union [7]. The eIDAS Regulation introduced the concepts of qualified trust service provider and qualified electronic time stamp and notes that "A qualified electronic time stamp issued in one Member State shall be recognised as a qualified electronic time stamp in all Member States" (Article 41/3).

3   Blockchain-Based Notary Services

In order to obtain a scientific, transparent and as complete as possible list of blockchain-based solutions to be used in this review, multiple searches were performed. The main search engine that was used was Google Scholar and during the final months of 2019. The adopted search terms that produced the most relevant results were "blockchain notary" and "distributed ledger notary". This selection was later broaden using the references comprises in the articles that were found on Google Scholar. Additionally, the generic Google search engine was also used to identify existing solutions that could possibly be already in use. The results of this searches are summarised next.

Existing European Solutions

Just recently, Slovenia has launched a national blockchain infrastructure to enable the deployment and testing of blockchain applications for the public and private sector. It was named Sl-Chain and is implemented on top of the distributed ledger named HashNET. In a press release, the Slovenian Government states that it was the first country, member of the European Union, to do so [18].

Luxembourg's Centre des Technologies de l'Information de l'Etat (CTIE), in conjunction with the Directorate-General for Informatics (DG DIGIT) of the European Commission, built a private blockchain network for notarisation [3]. It was implemented using Ethereum smart contracts, permits the registration of any type of documents using a data virtualisation layer. It assures integrity over time and time stamping. The project was public announced in march of 2019 [20].

In a press release [15] dated of 2017, the Italian notariat announced the release of the first blockchain for notary services in Europe. It was an exploratory project that allowed citizens, on a voluntary bases, to register content, in their *Notarchain*, to be shared with public institutions. The adopted blockchain model is the private/persmissioned model. Their aim was to develop a distributed system that assures unchangeability of data, but also accuracy and integrity. Additionally, it has preliminary support for identity check of involved parties.

Bernstein [6,13] is a Munich based company that offers some digital services similar to the existing trend of USA based digital notarisation services. In other words, their solutions enables proof of existence, integrity and time-stamping for digital documents. The documents can be stored on an InterPlanetary File System (IPFS). Their focus is on intellectual property assertion. OriginStamp [6] is also following the trend. It is a Switzerland-based company that offers time stamping with the help of the blockchain of the Bitcoin cryptocurrency. NotBot, the Blockchains Notary Bot, is yet another example. It is a time stamping service and its main purpose is to prove the existence of a document in a specific past date. It is operated by e-Genàse, a French company, and generates digital fingerprints of documents submitted to their NotBot online platform. These fingerprints are then stored on multiple blockchains (Bitcoin, NXT Blockchain).

The Swedish Mapping, Cadastral and Land Registration Authority, plus the Landshypotek Bank, Telia, ChromaWay, and others, developed a blockchain-based pilot to support real estate transactions and mortgage deeds registration [3,17]. Their solution adopts a private and permissioned blockchain that stores transaction information anonymously.

Table 1. Overview of European Notarisation services using blockchain

Solution	Country	Acts	Accessbility	Date	Blockchain
Sl-Chain	Slovenia	None	Private	2019	HashNET
DG DIGIT	Luxembourg	Time stamping, Integrity	Public	2019	Ethereum
OriginStamp	Switzerland	Time stamping, Integrity	Public	2019	Bitcoin
Notarchain	Italy	Integrity, Authenticity	Private	2017	Hyperledger Fabric
Bernstein	Germany	Time stamping, Integrity	Public	2016	Bitcoin
ChromaWay	Sweden	Land titleregistration	Private	2016	Postchain
NotBot	France	Time stamping, Integrity	Public	2016	Bitcoin, NXT

Table 1 summarises the identifies European blockchain-based solutions for notary services. Exception made for the Slovenian and Swedish solutions, the acts that can be supported be the European solutions are essentially document times tamping and its integrity validation. The Slovenian solution consists of a blockchain infrastructure as a service, crated with the ultimate purpose of leveraging the adoption of blockchain within notary services. The Swedish solution, following another blockchain-based world trend, is focused on supporting land title registration and transaction.

Rest of the World

Blocknotary [6], based in the USA, offers several services that revolve around the use of blockchain as a legal proof of authenticity. In fact, Act 157 of the State of Vermont's H.868, that went in effect on July 2016, includes a section regarding the use of blockchain for that purpose. Currently, multiple analogous service offerings can be found online. Blocksign [13] is a New York based service that offers a digitally signing service for digital documents that makes use of the blockchain of the Bitcoin cryptocurrency to store an hash of the signed document. Similarly, Bitcoin.com Notary [6,13] solutions uses the blockchain of the Bitcoin Cash (BCH) cryptocurrency to also enable document signing, time-stamping and verification. Acronis Notary [13] is yet another USA based blockchain service for document signing and verification, this one using the Ethereum blockchain. The Stampd [13] service only differs by its support for multiple back-end blockchains, including Bitcoin, Bitcoin Cash, Ethereum or Dash. Similarly, Stampery.com [6,13] also supports more than one blockchain for the storage of digital proofs, namely Bitcoin and Ethereum.

Mainly due to the global reach and immediate availability of multiple blockchains associated to cryptocurrencies, similar solutions can be found in parts of the world other than Europe or the USA. Proofstack (formerly CopyRobo), a Singapore based company, offers blockchain-based timestamping services that can be used as legal proof within Europe (eIDAS Regulation). Signatura [6], a company based in Argentina that also offers document signing and verification using blockchain, is such an example. Another example from South America was the partnership involving the Republic of Honduras and Factom to use blockchain to create a decentralised land title registry system [9,21]. This example demonstrates that legal and political issues may also hamper this type of solutions. Currently, Factom CEO has publicly stated that their Honduras project is stalled, until legislation can catch up with blockchain technologies [23]. The Republic of Georgia was yet another effort to support a land title registry system using blockchain technology [21].

Proof of Existence, ProveBit and Bitnotar [13] are open source solutions available online that support the storing of digital proofs on the Bitcoin blockchain. All three solutions have their respective source code available on GitHub. Proof of Existence, launched in 2013, is of particular note due to being the first readily available solution to use Bitcoin to verify that a document existed on a certain point in time.

4 Open Issues

Multiple issues can be raised when considering the applicability of blockchain technologies to the role of the notary as seen by the legal framework of the majority of the countries within Europe [2,8,10,12]. This are presented next and comprise: accuracy, reliability, authenticity, persistence, uniqueness, integrity, undemocratic operation, anonymity, resource usage and legal security.

When considering document **accuracy and reliability**, blockchain adds no material improvement [10]. While it may maintain an integrity representation of information (hash), it has no relation the records creation. The way records are created is the key determinant for its accuracy and reliability. One can easily argue that, even while storing hashes on a chain, if the initial records are erroneous, they will be referenced as valid by the blockchain.

The **authenticity** of records is of paramount importance for notaries, even being one of the services they provide. A human notary, while creating a record, it will verify not only the content of the record, but also the identity of the intervening persons and their legal capability to perform the act (being the owner of a land in a sale). Because this information is not stored within blockchain and due to the fact that transaction information is insufficient to identify the purpose of the data and its real world effects, the ascertainment provided by blockchain is limited [10].

Persistence through time is yet another issue. This one being a more consensual one [2,8,10]. Long term preservation of records in a fully distributed and open blockchain is achieved if it has sufficient nodes, if they are indeed distributed and if they maintain interest in participating in the consensus algorithm throughout time. No future assurances can be made regarding this participation [2,11].

Assuring the **uniqueness** of a record is somewhat cumbersome with current blockchain technologies [12]. One can easily change metadata or even a very small portion of the document to have it produce a distinct hash value. A different hash value will result in its interpretation as a distinct record by the system.

Integrity, or record immutability, is one of the recurrent adjectives associated to blockchain. But this is only true if specific conditions are met. In [2], the authors address the problem pose by bitcoin mining pools and their hash rate capabilities. This aspect is or particular concern if we analyse what is depicted in Fig. 1. It shows the bitcoin mining pool distribution[1], coloured by nationality, and one can see that $78,9\%$ (in red) are from the same country, China. Meaning that China is able to perform a structural majority attack on bitcoin (51% attack).

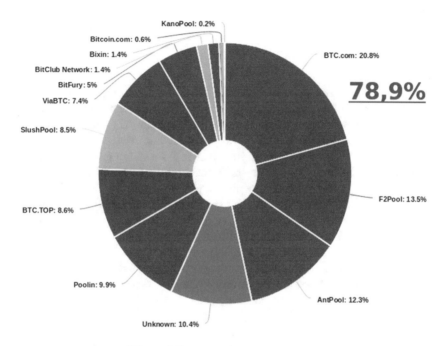

Fig. 1. Mining pools per nationality

The strength behind blockchain's consensus algorithms is achieved either with new block generation power or by possession of large quantity of stake. In both cases, if these are in control of a reduced set of individuals or groups, the system becomes **undemocratic**. The lack of individual and organisational responsibility in the operational control of the system [2] is yet another side effect.

The adoption of blockchain technologies that make use of pseudo-identities, such as wallet addresses, makes it highly suitable for fraudsters (tax fraud, money laundering, terrorist financing) [2]. The **anonymity** behind records may favour less legal behaviour.

Other issues related to **resource usage** such as storage space and the requires electricity to maintain a proof of work consensus algorithm are also

---

[1] Obtained from https://www.blockchain.com/pools in 30 of July of 2019.

identified in [2]. If one considers that blockchain stores references to records and not the records themselves, extra storage will be required outside of the protection umbrella of the blockchain. Another problematic legal issue may arise from users being lockout of their records by simply **losing cryptographic keys**.

In short, e-government cannot be represented by fully decentralised blockchains, despite some potential benefits [1]. Moreover, a trusted third party is still required in many use cases, even if blockchain-based technology is used [12]. Some authors [2] even conclude that blockchain technology is useful only for machine-to-machine communications.

5   Analysis

The use of blockchain technology in notarial activity can serve to ensure the authenticity of the sequence of acts related to land or property transmission by validating the supporting contracts, more so if these contracts are in digital form. A simplistic solution for the purpose of verifying the authenticity of the digital contracts supporting land or property transmission would be to calculate the hash of each document at the time of its archiving, and then store it in a blockchain in order to assure its immutability. In addition, to guarantee the authenticity of the contract documents, these would be digitally signed by the notary with his professional digital certificate. This digital signature would also serve as proof of the manifestation of the will of the involved parties in the performing the contract because the used electronic signatures are legally recognised. The Swedish solution described in Sect. 3 uses this approach.

Use of blockchain technology to support the activity of Common Law notaries, the most limited intervention of the notaries, can be achieved. The notary is replaced with the a blockchain system associated with digital signature schemes to the extent that they only act as a qualified witnesses and guarantee the authenticity of documents. As seen in Sect. 3 several solutions exist that are similar to these approach.

As far as Civil Law notaries operation is concerned, a blockchain contracting system can only assume an instrumental nature. However, it should be stressed that the use of blockchain could allow them to considerably improve the services they provide. Particularly by introducing a technological element that would enforce the transparency of the processes used to guarantee the immutability of the contracts in their custody. It could also permit that the evidential chain of ownership of assets could be audited by third parties, thus adding to the legal evidential value attributed by law to the acts of notaries.

6   Conclusion

There is a clear distinction between the legal role of a human Notary depending on jurisdiction. Common Law notaries perform a simpler role in society than their counterpart in Europe, Civil Law notaries. This lead to the quick appearance of blockchain-base solutions supporting the Common Law notary, but not the Civil Law notary. On one hand, due to nonexistent legal support, we can conclude that, within Europe, aside from specific use cases that include long term archival of documents or time stamping, it is currently not possible to avoid the use of human notaries. On the other, human notaries only have to gain with the adoption of transparency promoting technologies.

Worthy of note are the Italian legal initiative, that pushes for the adoption of blockchain-based solutions, and the Slovenian initiative, in creating a blockchain infrastructure for distributed application test and deployment.

References

1. Atzori, M.: Blockchain technology and decentralized governance: is the state still necessary? J. Govern. Regul. (2017). https://doi.org/10.22495/jgr-v6-i1-p5
2. Barbieri, M., Gassen, D.: Blockchain - can this new technology really revolutionize the land registry system. In: Land and Poverty World Bank Conference 2017: Responsible Land Governance (2017)
3. Bellia, M., et al.: Blockchain now and tomorrow - Assessing multidimensional impacts of distributed ledger technologies. Technical report, Joint Research Centre (European Commission) (2019)
4. Bracciali, A.: Italia il primo Paese a riconoscere validità giuridica della notarizzazione attraverso le DLT e gli Smart Contract? (2019). http://www. ilblogdellestelle.it/2019/02/italia-il-primo-paese-a-riconoscere-validitagiuridica-della-notarizzazione-attraverso-le-dlt-e-gli-smart-contract.html. Accessed 28 Jan 2020
5. Crosby, M., et al.: Blockchain technology: Beyond bitcoin. In: Applied Innovation 2.6-10, p. 71 (2016)
6. De La Rosa, J., Gibociv, D.: A survey of blockchain technologies for open innovation. In: EasyChair Preprint 830 (2019)
7. Dumortier, J.: Regulation (EU) No 910/2014 on electronic identification and trust services for electronic transactions in the internal market (eIDAS Regulation). In: EU Regulation of E-Commerce. Edward Elgar Publishing (2017)
8. DuPont, Q., Maurer, B.: Ledgers and law in the blockchain. Kings Review 23 (2015)
9. Feng, Q., et al.: A survey on privacy protection in blockchain system. J. Network Comput. Appl. **126**, 45–58 (2019)
10. Lemieux, V.: Blockchain and distributed ledgers as trusted recordkeeping systems: an archival theoretic evaluation framework, November 2017
11. Lin, I.C., Liao, T.-C.: A survey of blockchain security issues and challenges. IJ Network Secur. **19**(5), 653–659 (2017)
12. Locher, T., Obermeier, S., Pignolet, Y.A.: When can a distributed ledger replace a trusted third party? In: 2018 IEEE International Conference on Internet of Things (iThings) and IEEE Green Computing and Communications (GreenCom) and IEEE Cyber, Physical and Social Computing (CPSCom) and IEEE Smart Data (SmartData), pp. 1069–1077. IEEE (2018)
13. Di Francesco Maesa, D., Mori, P.: Blockchain 3.0 applications survey. J. Parallel Distrib. Comput. **138**, 99–114 (2020). https://doi.org/10.1016/j.jpdc.2019.12.019. http://www.sciencedirect.com/science/article/pii/S0743731519308664. ISSN 0743-7315

14. Nakamoto, S.: Bitcoin: a peer-to-peer electronic cash system (2008). http://bitcoin. org/bitcoin.pdf
15. Notaries of Europe: Italian notariat launches first notarial blockchain: "Notar-Chain". Notaries of Europe (2017). http://www.2017.notariesofeuropereport. eu/en/corner/italy-italian-notariat-launches-rst-notarial-blockchain-notarchain. Accessed 28 Jan 2020
16. Rocha, J.: Manual Teórico e Prático do Notariado. Edições Almedina (2003). ISBN 9789724015637
17. Salmeling, M., Fransson, C.: The land registry in the Blockchain - Testbed. A development project with Lantmäteriet, Landshypotek Bank, SBAB, Telia company, ChromaWay and Kairos Future (2017). http://chromaway.com/papers/ lockchainLandregistryReport2017.pdf
18. Republic of Slovenia: Slovenia launches national test blockchain infrastructure and Slovenian Blockchain partnership (2019). www.gov.si/en/news/slovenia-launches-national-test-blockchain-infrastructure-andslovenian-blockchain-partnership. Accessed 28 Jan 2020
19. Swan, M.: Blockchain: Blueprint for a New Economy. O'Reilly Media Inc., Sebastopol (2015)
20. Telindus: Luxembourg Notary Blockchain Kickoff – A first in Europe (2019). www.telindus.lu/en/blog/solutions-nance/luxembourg-notaryblockchain-kicko-rst-europe. Accessed 28 Jan 2020
21. Underwood, S.: Blockchain beyond Bitcoin. Commun. ACM **59**, 15–17 (2016). https://doi.org/10.1145/2994581
22. Wüst, K., Gervais, A.: Do you need a Blockchain? In: 2018 Crypto Valley Conference on Blockchain Technology (CVCBT), pp. 45–54. IEEE (2018)
23. Young, J.: Factom stalls honduras land title registry initiative (2019). www. newsbtc.com/2015/12/26/factom-stalls-honduras-land-titleregistry-initiative/. Accessed 28 Jan 2020

Qualified Targeting Through Data Aggregators in Permissioned Blockchain Settings: A Model for Auditable Transactions

Miguel-Angel Sicilia[1]($^{(\boxtimes)}$), Pedro Garrido[1], Salvador Sánchez-Alonso[1],
Marçal Mora-Cantallops[1], Elena García-Barriocanal[1], Salvador Casquero[2],
Lino González[1], and Alberto Ballesteros[1]

[1] University of Alcalá, Alcalá de Henares, Madrid, Spain
{msicilia,pedrojose.garrido,salvador.sanchez,
marcal.mora,elena.garciab,lino.gonzalez,alberto.ballesterosr}@uah.es
[2] 2gether, Crypto Plaza, C/. Don Ramón de la Cruz, 38, 28001 Madrid, Spain
salvador.casquero@2gether.global

Abstract. Data aggregators can leverage personal data for the benefit of their users in their role of mediators with enterprises that are willing to target qualified user segments with customized offerings. However, such type of transaction must comply with personal data protection regulation and provide a fair and auditable context for business partners. In this paper, the components of a solution for that problem in the context of a permissioned, private business network are described, along with an implementation using the Quorum enterprise derivative of Ethereum. The results of that design are expressed as a combination of schemas, protocols and smart contracts. The resulting model may serve as a blueprint for experimenting data market mechanisms that exploit the unique value of aggregators in a secure, transparent and auditable framework.

Keywords: Permissioned blockchains · User targeting · Personal data · Data aggregators · Smart contracts · Data markets

1 Introduction

Data aggregators play an important role in serving their users by consolidating their data in a single view. For example, financial account aggregators are able to collect data from multiple on-line bank accounts or other financial sites and provide services that make use or eventually substitute them as the preferred point of interaction [8]. One of those services is that of receiving tailored offerings, advertisements or proposals from third parties that are willing to target very specific segments of potential prospects. This may represent an alternative qualified channel to on-line advertising in Web search engines, social networks

J. Prieto et al. (Eds.): BLOCKCHAIN 2020, AISC 1238, pp. 111–120, 2020.
https://doi.org/10.1007/978-3-030-52535-4_12

or content platforms, that typically use real-time auctions for advertising and reward users "implicitly" by the provision of services [16].

However, data protection regulations, ethical concerns and potential user "reactance" [13] prevent that kind of interaction to happen with no restrictions. Further, from the viewpoint of companies originating the offerings, there is no easy way of having proof that the contacts have been made by the aggregator since there is no disclosure of personal data of those contacted. This complicates the scenario if a fair and transparent context is demanded by the incumbent third parties. More concretely, there are a number of requirements that need to be accounted for, including: (a) zero transfer of data from the aggregator to third parties to address personal data protection regulations [6,11], (b) a language that can be updated or extended for expressing customer segments, (c) auditable records of explicit user consent for being contacted, (d) avoid contacting prospects that are already customers of the third party company, (e) some mechanism enabling auditing that the right contacts have been done, and (f) a degree of fairness in accepting bids for targeting customers in the network of business contacts. This set of requirements together pose a significant challenge to the design of the system, beyond simple uses of a blockchain-centric, smart contract-based architecture as is common in current "Dapps" (e.g. those deployed on the Ethereum network). An additional problem, not discussed here in depth, is that of data pricing (as related to the price of addressing different customer segments), which should in principle account for profile rarity demand and type of offering among another factors.

Blockchain technologies appear as a potential candidate to support at least partially some of the above described problems. However, existing decentralized approaches to lead markets do not account for the these requirements or do not consider them in their value proposition. For example, *LeadCoin*[1] departs from a very different scenario, in which, according to their whitepaper "businesses sell their unused leads to other businesses and get value in the form of Lead-Coin tokens (LDC)." This entails that a seller is an interested party and not a neutral broker, and it resells personal information, so breaking (a). Also, leads are "burnt" when they stay long in the network, as they may have been resold many times. Further, it requires a mechanism of escrow for dispute resolution.

The design presented here makes the (strong) assumption that the aggregator is a neutral broker that benefits from the use of the network itself and not from particular transactions, so it does not have an incentive to manipulate prices or introduce bias. We discuss how the requirements described can be translated in concrete design elements identifying trust relations [14] and deployed in a permissioned environment, that fits the context of business collaboration in which an account aggregator is able to play a mediator role. The ideas described are preliminary and are intended only as a baseline towards resolving the mentioned challenges.

The rest of this paper is structured as follows. Section 2 describes the setting and case, and discusses the main data components identified in the blockchain setting. Section 3 describes the processes and relates them to the requirements of the situation. Finally, conclusions and outlook are provided in Sect. 4.

---

[1] https://www.leadcoin.network/.

2   Overall Setting

The architecture of the solution presented has three major subsystems. The off-chain subsystem is that of the aggregator, a conventional centralized system that holds all the information regarding users (and is thus subject to bilateral personal data regulation enforcement). The permissioned blockchain is the environment in which business partners (other companies) connect with the aggregator in a context of partial trust. The third subsystem is a Decentralized Identity (DID) deployment through which the users interact using P2P end-to-end secure transactions with the aggregator and eventually with third parties (using for example mobile apps), as those provided by the combination of Hyperledger projects Indy and Aries[2]. Those environments are isolated from each other and the aggregator mediates the delivery of offerings to users according to the consent given by them in an explicit way.

The preconditions for the interaction require a language for expressing user segments, an auditable way by which users give consent to be contacted, and a mechanism to identify contacts that are not already known by a given marketeer. These are discussed in the following sections.

The overall, high-level architecture is depicted in Fig. 1. In the bottom-right, the P2P secure interaction represents a completely separate network in which eventual interactions of data subjects with legal entities take place, initated by the former. The concrete cases in which that happens are described in the rest of this paper. The left part of the picture shows the Quorum network where the aggregator and the marketeers interact with the support of a number of smart contracts (dashed boxes represent the use of private bilateral transactions). It is important to know that the rewards to users are deliberately out of the picture. These should happen as external transactions between the aggregator and data subjects, either via traditional payment methods or using some cryptocurrency. In the latter case, the addresses of users in those cryptocurrency networks should not be disclosed to marketeers to avoid risk of linking payments to events in the permissioned network.

2.1   Profile Expression Language

Marketeers need to express conditions about user profiles to define user segments. This requires some form of declarative language for doing so and a schema. The requirement for such a language is that it is expressive enough to retrieve sets of users based on complex conditions. A simple and portable choice would be that of using SQL as the language, and the schema could be a regular relational schema. A Datalog dialect could be an alternative if logic programming features [2] are required or convenient, which is specially suited if the data has some triple structure as that used in the Web of Linked Data [5]. In any case, this language is to be executed (with or without transformation) on the aggregator system and not on the blockchain, so it can be changed with no impact on the

---

[2] https://www.hyperledger.org/projects/hyperledger-indy.

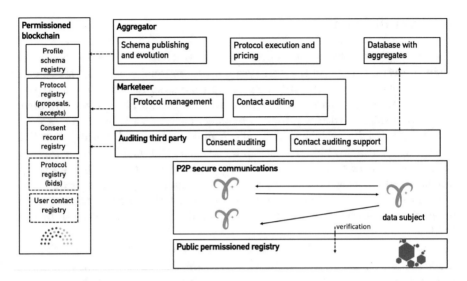

Fig. 1. High level depiction of the building blocks of the proposed solution

latter. The following would be an example of a Datalog-style query including a rule `has-car-insurance` that would entail traversal of facts in the database, and the clause `since` to restrict the time span of the query. That kind of time-based predicate can be found as built-in in databases as Datomic[3].

```
(d/q ('[:find ?user-id
:where
[?user-id :profile/age ?age]
[?user-id :hist/cash-advance-freq ?caf]
(has-car-insurance ?user_id)
(<= ?age 60)
(<= ?caf 10)
])
db since-2018)
```

The current schema can be published by the aggregator as a text or a pointer to a persistent file in a simple *ProfileSchemaRegistry* smart contract visible to all the parties. By default, the latest version of the schema will be retrieved, but all the previous versions are also registerd, allowing for evolution. Segments expressed by marketeers should end up in selections of a prototypical `User` table, entity or relation. Some proposed Ethereum standards might be used for such registry, as the *Ethereum Claims Registry* (ERC780) that allow to register arbitrary data with `setSelfClaim`, but for the kind of relations and functionality in our setting, a generic and standardized solution as that is not needed.

---

[3] https://www.datomic.com/.

2.2   User Consent Records

Users are able to consent on being targeted as results of queries for user segments, and they can also express conditions on the parts of the data they want to make available and the expiration date for such consent. It should be noted that this has implications in querying, as queries shall in principle discard users that have not consented in the use of a given field or piece of information that is in the given query, as not doing so entails non-matching uses to actually be matched due to implicit hiding of information.

They may also regulate the volume of communication allowed, and even more fine-grained tailoring as for example avoiding contacts at particular times of the week if needed. These requires some kind of *User Consent Record* (UCR) with the following main elements at a minimum:

- An opaque pseudonymous identifier of the user.
- An expression of the requirements, scope and restrictions of the consent.
- The digital signature of the user for the above information.

The aggregator would submit those UCRs to a *ConsentRecordRegistry* smart contract that would also allow for revoking the consent at any time. The identifier would be generated by the aggregator to link the UCR to its internal database. Those records will be encrypted by the aggregator, as they will be only made available to regulatory audits. A similar approach is that of *Chainscript* [1]. In our case, the records are maintained by the network to prevent tampering by the aggregator, who may eventually be accountable for proving that the use of data was according to consent. But it would be maintained encrypted by the aggregator so that it is not shared to other business partners, as it would violate requirement (a). It should be noted that the actual transfer and consent that is auditable by data protection agents lies in the DID subsystem.

User consent may also express consent for particular kinds of offerings or even companies using them. This would require having also available schemas describing so (as the in progress *Consent Receipt Specification* developed by the Kantara Initiative[4]), but this is omitted here for simplicity. In any case, these would act as additional filters for consent, as it fits other use cases [7].

2.3   Claims of Knowledge of Identity

One of the fundamental problems in targeted marketing via brokers is that of avoiding sending communications to users that are already customers of the brand for the given profile. This requires some form of communication between the aggregator and marketeers that must be one-directional, since no user data should flow to marketeers. A possible solution is that of using *claims of knowledge of identity* (CKIs) by which a marketeer shares a hash of a minimum piece of

---

[4] https://kantarainitiative.org/.

data (e.g. ID number + normalized name + date of birth) that the aggregator can match to know if a given user shall or not be contacted for that particular marketeer request.

The solution is not completely satisfactory as it effectively reveals personal information to the aggregator, that could be used to as a new fact at the aggregator's database. In this case, making the computation of the match private does not solve the problem, as the information is in the response itself. However, this assumes an adversarial setting of the aggregator and the marketeers, which we have disregarded as a departure assumption. Still marketeers may introduce noisy data so that perfect knowledge could not be attainable by a supposed malicious aggregator, but this is far from a completely satisfactory solution, and Zero-Knowledge-Proof (ZKP) technology does not appear to give a viable alternative since the proof of knowledge would effectively disclose personal data. Homomorphic encryption computing set intersections [3] could be an alternative but its implications and feasibility are not discussed here.

These CKIs should be interchanged via blockchain but in a private setting, as it is available for example in private smart contracts in the Quorum permissioned blockchain[5] or private channels in Hyperledger Fabric. This would avoid exposure to other potentially competing business partners while retaining bilateral auditability.

3   Processes

The elements described in the previous section account for requirements (b), (c) and (d) as stated in the introduction. They also do it with zero transfer of personal information to the marketeers. The requirements of fairness and auditing pose additional challenges to the design of the solution, as discussed in what follows.

3.1   Offering Bids and Responses

The process for a marketeer starts with composing a user profile segment expression using the language and schema provided by the *ProfileSchemaRegistry* contract. That expression should be packaged into a payload with the contents of the offering to be shown, its conditions and other configuration options, e.g. a simple case may be that of some HTML content to be displayed to targets with a link to a landing page and some expiration date. All that information would be part of a *private payload* that would be encrypted with the PK of the aggregator and if the size may be large, specified as reference to a distributed file system part of the blokchain deployment. In this case, a private channel or transaction is not an option, as there is other part of the bid that should be visible to other participants and linked to the private part for auditing.

The following are the steps of an interaction that occur through a *Targeting-Broker* contract.

---

[5]   https://www.goquorum.com/.

- **Bid**: Marketeer submits privately a transaction bid with: payload + bid price + number of targets desired (minimum and maximum).
- **Proposal**: Sent by the aggregator in response to a bid, including price that may or not match those of the bid.
- **Accept**: should include the funds materialized in terms of, for example, a utility token, or, depending on the deployment, some cryptocurrency.

When responding to bids, the aggregator has to first check CKIs relative to that transaction and then compute the users that are included in the segment requested minus collisions with CKIs. Then, it may create a proposal if it is viable when checking the UCR that have not expired. Expiration times for bids and proposals are required as a mean of automatically discarding interactions that for some reason become stalled.

Once a proposal is considered resolvable and interesting, the aggregator needs to set a price, which may be that offered by the corresponding bid or not. Since the aggregator is assumed to be neutral, bid price can be lower, and it should in all cases grounded on some mechanisms that could be auditable, as discussed in the next sub-section.

In order for other marketeers to avoid gaining information on the activity of their competitors, a proposal may include in the private payload a flag indicating a bogus bid (crafted by a marketeer as a way to generate noise), so that it should be disregarded automatically by the aggregator. This allows for transparent selection of bids in a context in which all business partners are able to observe proposals from the aggregator.

The processes described could be described in a business-oriented and domain specific language [9] for clarity to complement the actual code of the smart contracts, and to convey the steps that are actually being done off-chain by the aggregator and the marketeers.

3.2   Approaches to Pricing

The pricing as response to bids is the cornerstone of the incentive mechanism. The mechanism must strive to reach a balance of auditability – critical towards requirements (e) and (f) — and confidenciality in an environment of partial trust among business partners. Also incentives of users to share are critical [17] but these are not discussed here.

The idea is that the proposals for bids are signals that all business partners may use as information for their bidding requests, as they are visible to the network. However, that information is partial, as it includes time and amount but not the user segment expression, that can't be shared since it would reveal information to competitors. That partial information should be supplemented by the publication in the blockchain of key facts of the database of the aggregator[6]. Those facts may at least include the number of users but may also include other

---

[6] Note that this would require that the aggregator has a database that is immutable, i.e. an accrual of facts in time, as for example provided by Datomic.

facts, e.g. geolocation of customers. In doing so, the aggregator should account for methods of calculating statistical risk of data sharing.

The environment for the pricing is then a competitive one in which marketeers have only partial information, and the aggregator has a requirement of auditabilty. The approach for the proposals of the aggregator then needs to be based on some form of reinforcement learning in which the reward to be optimized is a notion of network-wide benefit encompassing at least three elements:

– Maximizing usage, which is in the interest of the aggregator as a mediator in the network. This has implications if the network has its own token and there are mechanisms by which value is transfered with use.
– Maximizing rewards to users according to profile rarity.
– Making the interchange efficient for marketeers, so that the inherent information asymmetry of the broker can eventually be proved not to be exploited by the aggregator.

Reinforcement learning has been applied to dynamic pricing with an inclusion of fairness concerns [10] however, the optimization criteria in our case is more complex, as it needs to balance some overall long-term goal with the interest of two different kinds of participants.

The design of an off-chain decision mechanism balancing the three aspects is not trivial and deserves separate attention.

3.3   Auditing Contacts

The acceptance of a proposal should trigger the sending of a communication or ad display to the selected users by the aggregator (on behalf of the marketeer that accepted the proposal). This has some privacy risk if the acceptance of the proposal is immediately sent, which may cause indirect disclosure of personal information, as the marketeer might attempt to correlate the acceptance of the proposal with actual hits in the landing page sent to users. Thus, the aggregator should introduce some stochasticity in the sending process, and finally registering for auditing a *Claim of User Contact* (CUC) as a private transaction with the marketeer. Those CUCs may include the timestamp and the links or references (e.g. hashes) to the originating bid and proposal, and the details of the contacts, being these encrypted with the PK of the aggregator to preserve privacy, or using opaque identifiers for which only the aggregator is able to establish a link with user information (different from those appearing in UCRs).

The problem of asymmetry here is that marketeers do not have ways to check that the right users have been contacted (also transparency may be important for users [4] and similar mechanisms may be used). To alleviate that problem, a marketeer may start an auditing event by selecting some random sample of the contacts[7] in its CUC registry and trigger an offering for auditing to these users via a third trusted party that acts as an auditor.

---

[7] A better approach may be that of contacting users that have been surveyed and voluntarily disclosed they knew of the proposal via the aggregator.

The actual contact with users would require a P2P communication protocol as Hyperledger Aries, so that it is the users that start the communication if they wish with the marketeer, knowing its DID and having an invitation. The marketeer may then ask a particular user details about the received offering, effectively checking the particular user was contacted without knowing its identity. This approach would require some sort of incentives, economic or other, for the user to answer auditing requests, and prevent overuse by marketeers.

While this mechanism does not constitute a complete solution as user response is not guaranteed and requires a third party, it can serve the practical purpose of establishing some mechanism of oversight about the behaviour of the aggregator.

4   Conclusions and Outlook

Information aggregators integrate disparate data into consolidated user profiles that are highly valuable to marketeers. They can play the role of a neutral mediator for the uni-directional communication of offerings to users based on that data. However, for this model to be viable, a number of components are required. We have discussed how a combination of a permissioned blockchain with a user target language and a set of mechanisms are able to set up an bidding-style channel that can be tailored to user preferences and consent and is trustable and fair for marketeers to a certain extent.

The model described is just a high level blueprint that requires configuration and some sort of empirical experimentation to determine the right prices and mechanisms, not only for the actual targeting events, but also for auditing. It is expected that models would be experimented as part of entrepreurial activity [15]. The solution is provisional in the sharing of contact knowledge using CKCs, and the mechanisms described are also limited in that they do not make the properties of the aggregator's database auditable by marketeers, e.g. facts as simple as the number of users in the database or other statistical profiles. That could be resolved by off-chain auditing of claims of such properties, but a completely automated process may be desirable. Future work should further improve the design in light of the limitations identified, and study alternative designs that can be used to evaluate the incentive mechanisms at the different points needed in the architecture.

Another direction for future research is that of integrating external, trustable data for pricing mechanisms, based on population statistics [12]. However, these are nowadays aggregates and more detailed demographics would be needed to be useful in the fine-grained model described in this paper.

Acknowledgements. Project funded by (proyecto financiado por) FEDER/ Ministerio de Ciencia, Innovación y Universidades - Agencia Estatal de Investigación/ Proyecto RTC-2017-6779-7, "Uso de blockchain en una plataforma financiera basada en Big Data, Inteligencia Artificial, machine learning, algoritmos predictivos y herramientas de lenguaje natural para la contratación, gestión e intercambio de activos no financieros."

References

1. Benchoufi, M., Porcher, R. Ravaud, P.: Blockchain protocols in clinical trials: transparency and traceability of consent. F1000Research, 6 (2017)
2. Ceri, S., Gottlob, G., Tanca, L.: What you always wanted to know about Datalog (and never dared to ask). IEEE Trans. Knowl. Data Eng. **1**(1), 146–166 (1989)
3. Chen, H., Laine, K., Rindal, P.: Fast private set intersection from homomorphic encryption. In: Proceedings of the 2017 ACM SIGSAC Conference on Computer and Communications Security, pp. 1243–1255 (2017)
4. Ertemel, A.V.: Implications of blockchain technology on marketing. J. Int. Trade Logist. Law **4**(2), 35–44 (2018)
5. Fayyaz, N., Ullah, I., Khusro, S.: On the current state of linked open data: issues, challenges, and future directions. Int. J. Semant. Web Inf. Syst. (IJSWIS) **14**(4), 110–128 (2018)
6. Finck, M.: Blockchains and data protection in the European Union. Eur. Data Prot. Law Rev. **4**, 17 (2018)
7. García-Barriocanal, E., Sicilia, M.A., Sánchez-Alonso, S.: The case for ontologies in expressing decisions in decentralized energy systems. In: Research Conference on Metadata and Semantics Research, pp. 365–376. Springer, Cham (2018)
8. Haikel-Elsabeh, M., Nouet, S., Nayaradou, M.: How personal finance management influences consumers' motivations and behavior regarding online banking services. Commun. Strat. **103**, 15 (2016)
9. Lagos, N., Mos, A., Cortes-Cornax, M.: Towards semantically-aided domain specific business process modeling. Data Technol. Appl. **52**(4), 463–481 (2018)
10. Maestre, R., Duque, J., Rubio, A., Arévalo, J.: Reinforcement learning for fair dynamic pricing. In: Proceedings of SAI Intelligent Systems Conference, pp. 120–135. Springer, Cham (2018)
11. Munier, L., Kemball-Cook, A.: Blockchain and the general data protection regulation: reconciling protection and innovation. J. Secur. Oper. Custody **11**(2), 145–157 (2019)
12. Sicilia, M.A., Visvizi, A.: Blockchain and OECD data repositories: opportunities and policymaking implications. Library Hi Tech **37**(1), 30–42 (2019)
13. Tucker, C.E.: Social networks, personalized advertising, and privacy controls. J. Mark. Res. **51**(5), 546–562 (2014)
14. Wessling, F., Ehmke, C., Hesenius, M., Gruhn, V.: How much blockchain do you need? towards a concept for building hybrid DAPP architectures. In: 2018 IEEE/ACM 1st International Workshop on Emerging Trends in Software Engineering for Blockchain (WETSEB), pp. 44–47 (2018)
15. Wu, Y.J., Chen, S.C., Pan, C.I.: Entrepreneurship in the internet age: internet, entrepreneurs, and capital resources. Int. J. Semant. Web Inf. Syst. (IJSWIS) **15**(4), 21–30 (2019)
16. Yuan, S., Wang, J., Zhao, X.: Real-time bidding for online advertising: measurement and analysis. In: Proceedings of the Seventh International Workshop on Data Mining for Online Advertising, pp. 1–8 (2013)
17. Zhou, T.: Examining users' knowledge sharing behaviour in online health communities. Data Technol. Appl. **53**(4), 442–455 (2019)

A Brief Review of Database Solutions Used within Blockchain Platforms

Blaž Podgorelec$^{(\boxtimes)}$, Muhamed Turkanović, and Martina Šestak

Faculty of Electrical Engineering and Computer Science, University of Maribor,
Koroška cesta 46, 2000 Maribor, Slovenia
{blaz.podgorelec,muhamed.turkanovic,martina.sestak}@um.si

Abstract. The rise of blockchain technology has inspired changes in the architecture of modern applications in various domains (e.g., finances, transport, etc.). Moving from a centralized data storage paradigm to the distributed ledger technology without central data storage raises important challenges, mostly in terms of scalability, integrity, and privacy. Even though decentralization, each node within the blockchain network implements a storage mechanism to keep track of the blockchain states. In this paper, we perform a detailed study of selected blockchain platforms in order to discover which underlying database solutions they use to store blockchain states. Our results show that 13 out of 20 platforms use key-value stores (LevelDB or RocksDB), write-optimized storage solutions able to perform fast lookups due to their underlying indexing structure used. Moreover, there are also platforms, which make use of the benefits of relational and document-oriented databases.

Keywords: Blockchain · Blockchain platform · Blockchain state · Data storage · Key-value · Data privacy · LevelDB · RocksDB

1 Introduction

In recent years, blockchain technology has received extensive attention in academia, as well as the enterprise and development circles. During this time, numerous blockchain platforms have been developed [1]. Furthermore, various research articles have been published [2–4], which address the question, if blockchain technology and blockchain-based applications are needed in different use-cases and if the current database solutions are adequate to meet all use-case requirements. Thus, one might assume that blockchain platforms and database storage solutions are technically separated. However, based on the documentation of various blockchain platforms (e.g., Ethereum [5], Hyperledger Fabric [6], Corda [7]), we can determine that databases are vital components of almost all blockchain platforms. Unlike before, when databases were seen as a dedicated component, they are now used within the very "core" of these blockchain platforms. More precisely, databases are used as a distributed storage component of each blockchain node for storing the global state of the blockchain network.

J. Prieto et al. (Eds.): BLOCKCHAIN 2020, AISC 1238, pp. 121–130, 2020.
https://doi.org/10.1007/978-3-030-52535-4_13

To the best of our knowledge, there is no comprehensive review, which outlines databases used within blockchain platforms. The main contribution of the paper is to provide a study of database solutions, used within various blockchain platforms, and analyze their characteristics and differences between them.

The rest of the paper is organized as follows: the theoretical background behind blockchain state storage and related research are described in Sect. 2. In Sect. 3, the methodology used to collect the required information is presented. Section 4 presents the obtained research results, and discusses the characteristics of various databases relevant for storing blockchain states. We conclude the article by outlining our research results and suggesting some future work.

2   Background and Related Work

Blockchain can be defined as a cryptographically-secured transactional machine, in which a cryptographically signed transaction modifies the state [5]. Within a blockchain platform, databases are typically used as components that store the state written in the shared blockchain ledger. A blockchain state is a global truth about the overall information verified and stored by all participants (i.e., peer-to-peer nodes) of the blockchain network. This information is verified through the validation component, as well as a dedicated consensus protocol (i.e., consensus component), which is included in all blockchain nodes (i.e., network component). Nodes within the consensus protocol agree about the correct state after a transaction is performed. The cryptography component is used within the entire process to ensure the integrity and security of the data sealed in the blockchain ledger [8,9].

Concurrency control mechanisms available in modern database solutions are required to handle larger volumes of simultaneous transactions in the context of blockchain systems [10]. Various authors have mentioned scalability as one of the biggest technical challenges of modern blockchain platforms. The sources of this challenge stem from the limited block size and limitations on transactions processed within a given time frame.

In [11], Tschorch and Scheuermann discuss how the scalability issue in the Bitcoin network has been handled by replacing the existing BerkeleyDB data storage solution with LevelDB. The authors concluded that the change increased performance of the ledger's synchronization and block verification.

Chen, Lv, and Song [12] tackle the storage-caused performance issues by constructing a novel data storage model suitable for the consortium blockchain, in which they combine blockchain and traditional data storage methods.

In most blockchain systems, the key-value or file system storage and code-level APIs can be found to store block data. Zhu et al. [13] propose to re-use existing database solutions' mechanisms. Their approach includes storing transactions into relational tables and querying these tables with a SQL-like language syntax.

A different category of blockchain platforms are those, which combine the properties of both blockchain and database systems, such as ChainSQL [8] and BigchainDB [14]. BigchainDB merges the decentralization and immutability properties of blockchain systems with high transaction rates, low latency, and querying structured data properties of database systems, in order to combine the best of the two technologies. In a BigchainDB 2.0 network, each node stores block data into its local MongoDB database, and decides through which mechanism the stored data will be available to others (REST API or GraphQL API) [14]. Compared to this approach, ChainSQL enriches the decentralized blockchain features with optimized query processing and data structures developed in distributed databases [8].

The distributed approach can also be found in [15], where only a subset of changes by transactions within a given block is stored into distributed hash tables, instead of storing the entire blockchain history. This approach helps to minimize the amount of data that must be stored in the validating node. The authors develop a distributed approach, which, unlike sharding, does not require a complete re-engineering of the blockchain, and the size of the stored values is not so dependent on the size of the shard.

Additionally, several authors discussed the topic of implementing blockchain data provenance. In [16], Wang et al. present a specialized storage engine for blockchain applications called Forkbase. The engine implements some necessary features for the distributed ledger environment, which are not available in modern database systems (e.g., explicit data versioning, tamper-evident ledger, resolving concurrent write access). On the other side, modern blockchain platforms, such as Ethereum, use additional key-value stores or file systems to implement the previously mentioned features. The authors argue that usage might bring additional development cost and performance overhead without any significant contribution to the scalability or analytical power of these systems [16].

In [10], Dinh et al. present a qualitative analysis of selected private and public blockchain systems with regards to the distributed ledger, cryptography, consensus protocol, and smart contract concepts. The authors did not direct their research efforts to the underlying storage solutions used in the analyzed blockchain platforms, nor to the provenance approaches for storing blockchain states. Instead, they analyzed the platforms only from the perspective of the underlying data model implemented for tracking database state (e.g., transaction- or account-based model). Conversely, our research aims to provide an overview of specific underlying database solutions used in modern blockchain platforms, which has not been performed so far in the research literature.

3   Methodology

At the moment, there are a variety of blockchain platforms developed. Thus, we initially had to determine which blockchain platforms are suitable to be included in our analysis. Because there is no list of all existing blockchain platforms available, we chose to follow the list published by Gartner, which includes primary blockchain platforms utilized in testing of the blockchain technology on various use-cases by different companies [17].

The complete selection process is presented in Fig. 1. The component at the top presents an initial list of all blockchain platforms used in the selection process. Each step is presented with a rectangle marked with the number. On the rounded rectangles at the left and right side, there are lists of blockchain platforms that were added (indicated by a plus sign) or removed (indicated by minus signs) upon the initial list. Finally, the component at the bottom presents a final list of blockchain platforms included in the analysis.

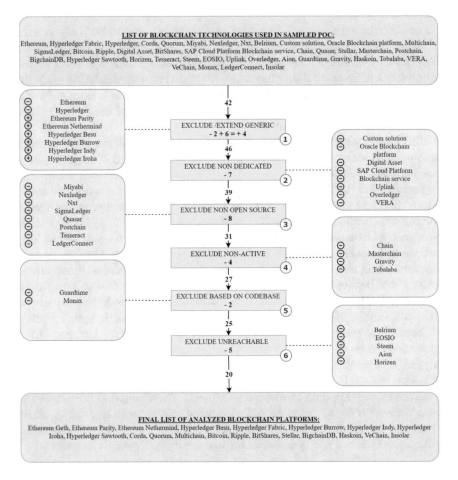

Fig. 1. Blockchain platforms selection process.

In the first step, we excluded the umbrella projects (i.e., Ethereum, Hyperledger) from Gartner's original list, since these include multiple platforms (e.g., HL Fabric or HL Iroha), and use different database solutions. Therefore, we expanded the list with the individual blockchain platforms (Ethereum Parity, Ethereum Nethermind, Hyperledger Besu, Hyperledger Burrow, Hyperledger Indy, Hyperledger Iroha).

In the second step, seven non-dedicated platforms were excluded. These platforms either: (1) are in the form of platform-as-a-service (Oracle Blockchain platform, SAP Cloud Platform Blockchain service), which offer the usage of dedicated platforms; (2) enable only specific features (e.g., smart contracts, Dapp, inter-blockchains connectors), which cannot be represented as the core of the blockchain platform (Digital Asset, VERA, Overledger), or (3) are not a blockchain platform at all (Custom solution, Uplink).

Next (i.e., third step), to be objective as possible, we chose to determine the database solution used by a blockchain platform, based on its codebase, since the documentation of the blockchain platforms rarely clearly defines which database solution is used within that platform. Due to such an approach, we excluded the platforms, which do not have an open-source codebase, the components cannot be verified. Such platforms are Miyabi, Nexledger, Nxt, SigmaLedger, Quasar, Postchain, Tesseract and LedgerCGuaronnect. Furthermore, in the fourth step, archived and not maintained blockchain platforms were excluded as well (i.e., Chain, Masterchain, Gravity and Tobalaba).

In the following step (i.e., fifth step), platforms Guardtime and Monax, for which we were unable to determine the underlying database solution from the codebase, were excluded. As a result of the last step, 25 platforms remained. For these platforms, the information about database solutions was obtained from their codebase. However, since this information is entirely based on the codebase, the results needed to be verified additionally.

To verify the results, we contacted the remained platform owners or active contributors to verify or correct our findings. As an outcome of this step (i.e., sixth step), the information about database solutions was successfully verified for 20 platforms. For the remaining five platforms, we did not receive any response to our inquiry, so we excluded these platforms from further research, as their information could not be verified. The final results, which contain verified information about the used database solutions within each blockchain platform, are presented in Table 1 in Sect. 4.

4   Results and Discussion

Overall, the results of our research presented in Fig. 2 show that 13 out of 20 blockchain platforms, for which the information could be collected, use either LevelDB [18] or RocksDB [19] databases to store blockchain state data. The decision to use key-value stores for this scenario is probably driven by the fact that storing blocks and other blockchain data in key-value pairs logically results in faster look-ups [20]. Nevertheless, platforms such as HyperLedger Fabric and Ripple use multiple options for storing blockchain states, where one of them is selected as the default store, and the other as an alternative external option to further extend the querying possibilities. In the following paragraph, we elaborate on the characteristics of a few selected database solutions, which might justify their usage in blockchain platforms. Due to content limitations, we elaborate only on LevelDB and RocksDB, due to the percentage involved in the blockchain platforms, as well as PostgreSQL and LMDB, due to their specific properties.

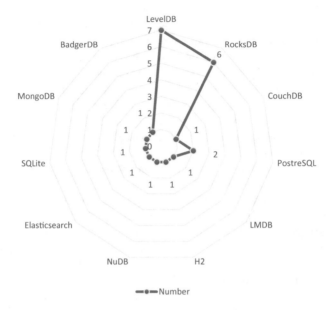

Fig. 2. A total number of individual database solutions used among analyzed blockchain platforms.

In its current implementation, a key-value NoSQL **LevelDB**, used by the majority of analyzed platforms, uses a write-optimized log-structured merge (LSM) tree, which significantly reflects in the query performance during large sequential writes (insertions or deletions) [18]. Conversely, the read performance is improved by automatically compressing data before persistently storing it to the disk. Also, caching is used to avoid decompressing data for each query, because repeatedly decompressing blocks read from the disk can result in high read costs. Aside from being owned and maintained by Google, one of the reasons for its wide usage is the support for most modern programming languages (C++, Go, Java, JavaScript, etc.) [21]. LevelDB supports both forward and backward iteration over data, where data is handled by calling put(), get(), delete() and batch() operations. Blockchain data is stored as ordered key-value pairs of string sequences (byte arrays in the background), which is a more native storage format for blockchain data. Database states are stored as read-only snapshots, which can be referenced when needed. Writing data in batches brings atomicity benefits (no changes in data are lost if the writing process crashes), as well as increased speed of bulk updates by placing many individual similar data mutations into a single batch [21]. The read performance can also be improved by grouping adjacent keys into the same block. Additional writing efficiency is achieved by

Table 1. Databases solutions used for state storage within blockchain platforms.

Project	Version	Database solution
Ethereum Geth	v1.9.10	LevelDB
Ethereum Parity	v2.6.8-beta	RocksDB
Ethereum Nethermind	v1.4.7	RocksDB
Hyperledger Besu	v1.4.0-beta2	RocksDB
Hyperledger Fabric	v2.0.0-beta	LevelDB/CouchDB
Hyperledger Burrow	v0.29.5	LevelDB
Hyperledger Indy	1.12.1	RocksDB
Hyperledger Iroha	v1.1.1	PostgreSQL
Hyperledger Sawtooth	v1.2.3	LMDB
Corda	release-os-4.4-RC01	H2
Quorum	v2.4.0	LevelDB
Multichain	v2.0.5	LevelDB
Bitcoin	v0.019.0.1	LevelDB
Ripple	v1.4.0	NuDB/RocksDB
BitShares	v3.3.2	Elasticsearch
Stellar	v12.2.0	SQl-based (SQLite/PostgreSQL)
BigchainDB	v2.0.0	MongoDB
Haskoin	v0.9.15	RocksDB
VeChain	v1.2.0	LevelDB
Insolar	v1.3.0	BadgerDB

configuring the block size according to needs. As a concurrency control mechanism, LevelDB uses two-phase locking (2PL), allowing only one process at a time to write data into the database, while multiple threads can access the data. LevelDB also offers a built-in function, which can be run to recover data as much as possible in case of database corruption, and it supports replication.

As a forked database solution, **RocksDB** shares many characteristics and features with LevelDB [19]. It was developed by Facebook to meet the high-performance requirements of key-value stores in the era of flash storage [22]. Hence, its primary goal is to fully use the fast access speed provided by flash storage, and adapt to different workloads. The underlying stored data structure does not differ much to LevelDB (ordered key-value pairs of byte streams), but RocksDB uses the column-oriented approach in persistent storage. Specifically, RocksDB uses the concept of column families (e.g., blocks, transactions) to partition the database [19] logically. This change results in atomic writes ensured for each column family, as well as provided support for flexible configuration settings to gain most benefits in different production environments (e.g., HDFS, flash storage or hard disk, pure memory, or other). However, the atomicity property is currently not guaranteed after database recovery for more than one column family [19]. Different to LevelDB, RocksDB supports both pessimistic (PCC) and optimistic (OCC) concurrency control mechanisms, since the efforts are directed towards improving write/read performance to the persistent storage instead of retaining much data inside the memory. This approach prevents additional storage costs caused by the increasing amount of RAM required by the increasing size of the blockchain. By using the OCC mechanism instead of 2PL, there is a need for locking data records for other users while only one user makes changes to it, and more users can connect to the server since fewer server resources are used on maintaining "idle" database connections. Additionally, RocksDB includes the support for defining indexes to speed up the data retrieval process. This presents a significant benefit for new nodes added to the blockchain or mining and validation nodes, which need to retrieve blockchain history (i.e., "download" data) in order to confirm transactions.

Unlike LevelDB or RocksDB, **LMDB**, used by Hyperledger Sawtooth, uses B+ trees to manage stored data, which results in better read performance. However, the differences in the design of LSM and B+ trees have a significant influence on read/write performance in larger workloads. A benchmark test performed by Dix in [23], in which LevelDB, RocksDB, and LMDB were tested on read/write operations on a varying number of values, showed that LMDB achieves best read performance on a smaller number of values. This is directly related to the underlying B+ tree implementation. In contrast, in larger workloads, LSM tree-based solutions perform better in both read and write operations. In its underlying storage model, LMDB uses a memory-map to store the database, so caching and other semantics depend on the operating system [24].

As a relational DBMS, **PostgreSQL**, used by Stellar and Hyperledger Iroha, requires a fixed table schema to be specified and followed. The fixed schema can bring some challenges and difficulties when storing blockchain data. Namely, database designers come across an issue of selecting the most efficient data type to store large hash values of blockchain transactions. Most often, these values are stored either as INTEGER or BIGINT data types; however, sometimes, this may not be the most efficient strategy for several reasons (e.g., the mismatch between 32- and 64-bit values can cause a significant amount of storage to remain unused) [25]. Most importantly, different schema design strategies, which heavily depend on database designer's skills, may have a significant impact on later query performances. For instance, some designers may introduce generic table identifiers in addition to already present transaction IDs, which will increase the query execution time during table joins. Also, backward iteration over data may not be so simple to implement, as this feature is not natively supported in PostgreSQL. Nevertheless, the Atomicity, Consistency, Isolation, Durability (ACID) integrity model behind PostgreSQL ensures that all data rows remain in the system even after they are updated or deleted [25], which contributes to blockchain data provenance. Using PostgreSQL also brings benefits in terms of data recovery and availability; namely, each database node stores blocks into separate stores, so if a given node were to have corrupt records at some time, the other replica nodes would remain intact.

5   Conclusion

Modern relational and non-relational database solutions offer different features to store blockchain state data. Nevertheless, our results indicate that 13 out of 20 analyzed blockchain platforms choose to logically store this data in key-value stores with varying underlying storage formats (relational or column-oriented) based on LSM tree index structure. Even though LevelDB and RocksDB might come with some write/read amplification due to their storage strategy, they still include a variety of other mechanisms and features. Their features can be used to mitigate this issue, and achieve a satisfactory read/write performance and throughput required by the platforms (e.g., caching, indexing, batch writes/reads). Besides key-value stores, the modularity and scalability of relational (e.g., PostgreSQL) and document-oriented databases (e.g., MongoDB) are also some additional features consumed by some platforms (e.g., BigChainDB, Hyperledge Iroha). These also incorporate positive features like complex querying support with SQL. As part of our future work, we plan to study and justify the usage of relational, NoSQL and NewSQL database solutions in different blockchain-based use cases, as well analyze the effects of using different indexing techniques and mechanisms to improve blockchain performance. We also plan to design a performance benchmark for blockchain-based systems, focusing on the underlying database storage features.

Acknowledgments. This research was funded by the Slovenian Research Agency (research core funding No. P2-0057), and also in parts by EC H2020 Project CON-CORDIA GA 830927.

References

1. Belotti, M., Božić, N., Pujolle, G., Secci, S.: A vademecum on blockchain technologies: when, which, and how. IEEE Commun. Surv. Tutorials **21**(4), 3796–3838 (2019)
2. Koens, T., Poll, E.: "What blockchain alternative do you need?" In: Data Privacy Management, Cryptocurrencies and Blockchain Technology, pp. 113–129, Springer, Cham (2018)
3. Wüst, K., Gervais, A.: Do you need a blockchain? In: 2018 Crypto Valley Conference on Blockchain Technology (CVCBT), pp. 45–54, IEEE (2018)
4. Peck, M.E.: Blockchain world-do you need a blockchain? this chart will tell you if the technology can solve your problem. IEEE Spectrum **54**(10), 38–60 (2017)
5. Wood, G., et al.: Ethereum: a secure decentralised generalised transaction ledger. Ethereum Project Yellow Paper **151**(2014), 1–32 (2014)
6. Cachin, C., et al.: Architecture of the hyperledger blockchain fabric. In: Workshop on Distributed Cryptocurrencies and Consensus Ledgers, vol. 310, p. 4 (2016)
7. Hearn, M.: Corda: a distributed ledger. Corda Technical White Paper, vol. 2016 (2016)
8. Muzammal, M., Qu, Q., Nasrulin, B.: Renovating blockchain with distributed databases: an open source system. Fut. Generation Comput. Syst. **90**, 105–117 (2019)

9. Zheng, Z., Xie, S., Dai, H., Chen, X., Wang, H.: An overview of blockchain technology: architecture, consensus, and future trends." In: 2017 IEEE International Congress on Big Data (BigData congress), pp. 557–564, IEEE (2017)
10. Dinh, T.T.A., Liu, R., Zhang, M., Chen, G., Ooi, B.C., Wang, J.: Untangling blockchain: a data processing view of blockchain systems. IEEE Trans. Knowl. Data Eng. **30**(7), 1366–1385 (2018)
11. Tschorsch, F., Scheuermann, B.: Bitcoin and beyond: a technical survey on decentralized digital currencies. IEEE Commun. Surv. Tutorials **18**(3), 2084–2123 (2016)
12. Chen, J., Lv, Z., Song, H.: Design of personnel big data management system based on blockchain. Fut. Generation Comput. Syst. **101**, 1122–1129 (2019)
13. Zhu, Y., Zhang, Z., Jin, C., Zhou, A., Yan, Y.: Sebdb: semantics empowered blockchain database. In: 2019 IEEE 35th International Conference on Data Engineering (ICDE), pp. 1820–1831, IEEE (2019)
14. McConaghy, T., Marques, R., Müller, A., De Jonghe, D., McConaghy, T., McMullen, G., Henderson, R., Bellemare, S., Granzotto, A.: Bigchaindb: a scalable blockchain database. White paper, BigChainDB (2016)
15. Bernardini, M., Pennino, D., Pizzonia, M.: Blockchains meet distributed hash tables: decoupling validation from state storage *arXiv preprint* arXiv:1904.01935 (2019)
16. Wang, S., Dinh, T.T.A., Lin, Q., Xie, Z., Zhang, M., Cai, Q., Chen, G., Ooi, B.C., Ruan, P.: Forkbase: an efficient storage engine for blockchain and forkable applications. Proc. VLDB Endowment **11**(10), 1137–1150 (2018)
17. Groombridge, D., Healey, C.: Blockchain trials show pragmatism emerging across industries. Accessed 22 Dec 2019
18. Dean, J., Ghemawat, S.: Leveldb github repository. https://github.com/google/leveldb/blob/master/doc/index.md. Accessed 14 June 2019
19. Siying, D.: Welcome to rocksdb. https://github.com/facebook/rocksdb/wiki. Accessed 21 Sept 2019
20. Riegger, C., Vinçon, T., Petrov, I.: Efficient data and indexing structure for blockchains in enterprise systems. Proceedings of the 20th International Conference on Information Integration and Web-based Applications & Services, pp. 173–182 (2018)
21. Group, C.M.D.: "Leveldb." https://dbdb.io/db/leveldb (2020)
22. Group, C.M.D.: "Rocksdb." https://dbdb.io/db/rocksdb (2020)
23. Dix, P.: Benchmarking leveldb vs. rocksdb vs. hyperleveldb vs. lmdb performance for influxdb. https://www.influxdata.com/blog/benchmarking-leveldb-vs-rocksdb-vs-hyperleveldb-vs-lmdb-performance-for-influxdb/ 20 June 2014
24. Group, C.M.D.: "Lmdb." https://dbdb.io/db/lmdb (2020)
25. Trubetskoy, G.: Blockchain in postgresql part 2. https://grisha.org/blog/2017/10/20/blockchain-in-postgresql-part-2/ 20 Oct 2017

A Framework for On-Demand Reporting of Cryptocurrency Ownership and Provenance

Rui Carreira[1], Pedro Pinto[2(✉)], and António Pinto[3,4]

[1] Instituto Politécnico de Viana do Castelo, 4900-347 Viana do Castelo, Portugal
`ruicarreira@ipvc.pt`
[2] Instituto Politécnico de Viana do Castelo, ISMAI and INESC TEC,
4900-347 Viana do Castelo, Portugal
`pedropinto@ipvc.pt`
[3] CRACS & INESC TEC, Porto, Portugal
`apinto@inesctec.pt`
[4] CIICESI, ESTG, Politécnico do Porto, Porto, Portugal

Abstract. Payments using cryptocurrencies may require that the user is able to provide proof of ownership and proof of provenance for a specific transaction. In this paper an innovative web based solution is proposed as a framework that issues reports, on request, pertaining proof of ownership and proof of provenance. The proposed framework provides proof of ownership by using micro-payments and, when used recursively, it can produce provenance reports up to a defined granularity level of transactions. A proof of concept prototype of the proposed framework was implemented and its operation and output is presented and explained. Some limitations and future work directions are also identified.

Keywords: Proof of ownership · Proof of provenance · Cryptocurrency · Blockchain

1 Introduction

Bitcoin is a digital currency, proposed in 2008 by Satoshi Nakamoto [1]. Bitcoin bypasses the centralized currency management and thus, it is a decentralised digital currency that uses an open, public and anonymous blockchain network, named Bitcoin blockchain, to store the transactions between cryptocurrency wallets.

The Bitcoin blockchain uses blocks to store data for each transaction, and each block includes the value for a given transaction, and the public addresses of

A. Pinto—This work is partially financed by National Funds through the Portuguese funding agency, FCT - Fundação para a Ciência e a Tecnologia, within project UIDB/50014/2020.

J. Prieto et al. (Eds.): BLOCKCHAIN 2020, AISC 1238, pp. 131–143, 2020.
https://doi.org/10.1007/978-3-030-52535-4_14

the source and destination wallets used in each transaction. These public wallets addresses, by themselves, can not be associated to the real identity of the wallet owner.

A cryptocurrency wallet, contains a pair of public and private cryptographic keys. The public key can be shared with anyone and allows other wallets to make payments to this wallet. The private key is not meant to be shared as it used to create a signature for the transaction and thus, it is used to prove that the transaction was made by the true wallet owner.

Specific payment services that use cryptocurrencies may require its owner to prove ownership and to demonstrate the origin of the cryptocurrency, i.e. to prove the provenance. These situations tend to occur in countries that impose anti-money laundering regulations but do not recognize cryptocurrencies as a legal form of money. For instance, in Portugal, if a citizen buys a property over €250.000, he is required to report it in his Personal Income Tax declaration (or in portuguese, "Imposto sobre o Rendimento de Pessoas Singulares"). If audited and requested, he must provide proof of ownership of the money used, and its proof of provenance, even if he bought the property using cryptocurrency (in Portugal, the cryptocurrencies do not have legal tender and they are not recognized as money).

This paper proposes a framework capable to provide the proof the ownership over one or more cryptocurrency wallets and also enables the proof provenance of particular values transacted between wallets. The proposed framework comprehends a service that relate specific owners to their wallets by using micro-payments and generates a wallet ownership report by request, which contains personal information about the owner (name, personal addresses and id numbers) and the wallet address. These procedures, used recursively, allows this framework to certify currency ownership and provenance, up to a defined granularity level.

This paper is organized as follows. The Sect. 2 reviews the current challenges regarding cryptocurrency proof of ownership and provenance. The Sect. 3 describes the proposed framework. Section 4 presents the operation and outputs of the implemented prototype. Section 5 concludes this paper, presents limitations of the current proposal and highlights possible directions for future work.

2   Review on Proof of Ownership and Provenance

According to [2], an electronic cash (e-cash) system acceptance and successful usage depends on a balance between anonymity and traceability. Traceable e-cash would make it harder to commit many crimes but would also hamper the privacy of users. Untraceable e-cash would facilitate crime and money laundering schemes. Ultimately, Gemmell argues that e-cash should, at least, be partially traceable. Bitcoin [1], if we consider the cryptocurrency revolution it caused and its worldwide adoption, seems to have achieved this balance by promoting full transaction traceability and pseudo-anonymity of the involved entities. The pseudo-anonymity is achieved by allowing each user to create one, or more, digital identities (or wallet addresses), and even exchange currency between them.

Meaning that, any person can use the cryptocurrency pseudo-anonymously, fostering its adoption without fear of being monitored by state or government agencies, but police forces may still promote investigations pertaining the use of cryptocurrencies. One must recall that, upon identification of the person behind a digital wallet, all transactions of such wallet can be easily identified and mapped to the person.

Nonetheless, and due to their decentralized nature, governments are not in control of key cryptocurrencies such as bitcoin or Ethereum. Some efforts have been made in this direction [3] with anti-money laundering regulations imposed to virtual currency exchanges that exchange cryptocurrency for fiat currency. Example being the Fifth Money Laundering Directive, published by the European Union (EU) in June of 2018, that explicitly addresses cryptocurrency regulation. Similar efforts are being made in the United States of America (USA). In a news report by Mizrahi, one can see interest in promoting the regulation of cryptocurrency Exchange Traded Funds (ETF) [4], but also that the Securities and Exchange Commission (SEC) is seeking for a service that will identify owners of cryptocurrencies.

Distinct forms of attestation that include proof of ownership, proof of existence, proof of integrity of documents or proof of receipt of some information, are possible due to the use of blockchain [5–7]. Proof of existence relates to the attestation that some digital document existed at a certain time. Proof of integrity relates to the attestation that some document was not tampered with. Proof of receipt relates to the attestation that someone received a certain information or document. When considering concepts such as proof of ownership or proof of provenance, these are not consensual. Proof of ownership can either relate to the attestation that someone authored a document [8] or it can relate to someone owning a specific wallet or a value in some cryptocurrency [9]. Proof of provenance, in the digital world, usually relates to data provenance, in other words, to prove that some data was originated in some other system or entity [10].

The concepts pertaining the work herein are then being able to prove wallet ownership and to prove the origin of the cryptocurrency in such wallets. These capabilities maybe required when someone wants to purchase a high value asset, such as a house or a luxury car, in countries that haven't recognized cryptocurrencies as a form of money. Moreover, if one combine this legal context with anti-money laundry regulations, the problem aggravates. For this reason, multiple online services appeared to offer services related to either the tracking of cryptocurrency between digital wallets [11], or to the proof of wallet ownership [12].

For instance, Coinfirm states that they provide Anti Money Laundering (AML) and Know Your Customer (KYC) compliant services to major financial institutions, in jurisdictions all over the world, including the UK, EU, Switzerland, Japan and Central & Eastern Europe. This company also issues Certificates and Reports to their costumers. Some of these certificates contain a proof of wallet ownership that can be confirmed in two ways, depending on the virtual currency, and consist in using a cryptographic signature or a micro payment. When using a cryptographic signature, the owner of wallet signs any text string using his private key and provides the investigator with an encrypted message, called a digital signature. The signature is then verified by the algorithm against the wallet address. When using micro payments, the investigator asks the owner of the wallet to send a micro-payment to the investigator's wallet address, which is immediately sent back to the owner's wallet. Additionally, investigator asks for a screen shot of the transactions. Together they are considered sufficient for verification process.

3   The Proposed Framework

The proposed framework is designed to provide users certificates of wallet ownership and reports of transactions on the associated wallets. It should operate as a proof generation service, in a cloud-based environment, with two types of roles, the user and the admin with the following requirements and interactions. In the system, the user should be able to associate one or more wallets to his account and then he should be able to request a wallet ownership report for the associated wallets. In order to confirm the association of the user account with a given wallet, the user should be required to make a micro-payment. The system will check whether the transaction was performed and the user is then notified if it was successful or not. In any case, whether the micro-payment has been performed or not, the system will be activated to allow the return of this micro-payment. In case the micro-payment is confirmed, the user should be able to view information of his wallets, to request a wallet ownership report from the system and check the current status of all the reports requested. The User Use Cases diagram is as depicted in Fig. 1.

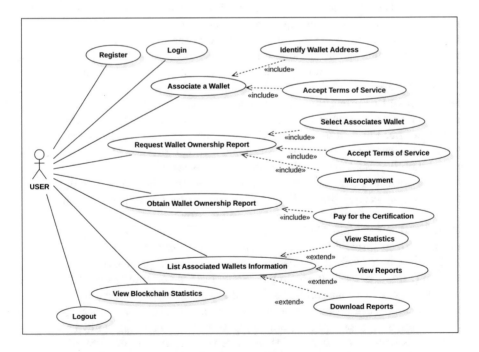

Fig. 1. User Use Cases diagram

The administrator should be able to login check all the micro-transactions and wallet ownership reports payments and validate them. The Administrator Use Cases diagram is as depicted in Fig. 2.

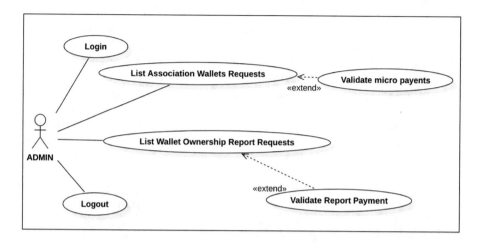

Fig. 2. Administrator Use Cases diagram

The architecture of the proposed framework is depicted in Fig. 3 and it comprises a front-end application, a back-end application and a database. The users access the framework and every interaction with the back-end and the database is secured by Passport.js [13] which manages the JSON Web Tokens [14].

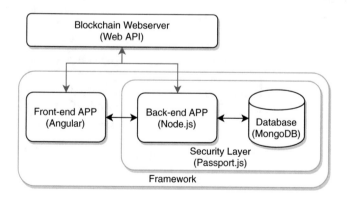

Fig. 3. Framework architecture

The front-end application is used by both users and administrators. The users can manage and check their wallets information, namely they can associate new wallets to their accounts and request a wallet ownership certification. The user dashboard is also consuming a web API to show blockchain-related information such us as the current bitcoin market price and statistics. The administrators can check all the information related with the users wallets ownership requests and respective payments. So, the administrators can manually verify if the payments are being executed and consequently emit the respective ownership certifications.

The back-end application is responsible for the management and the treatment of all the data shown on the front-end application and for the database feed. All the communication between the back-end and front-end application makes use of Json Web Tokens that prevents unauthorized access to certain functionalities of the system itself. The back-end application do also communicates with a Blockchain web API to access the users wallets transactions and information.

The Database stores the personal information of the users and their subscriptions, the associated wallets and respective transactions. The domain model of the database is presented in Fig. 4.

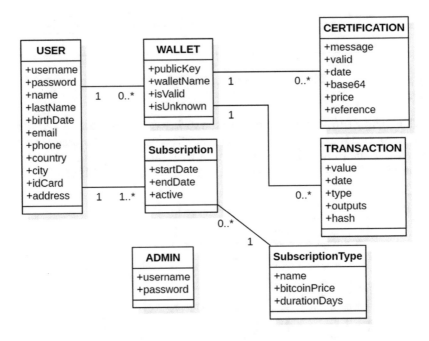

Fig. 4. Database

4   Framework Operation and Output

The framework operates and provides outputs for users and administrators, each with their interfaces and internal modules, as presented in the Fig. 5.

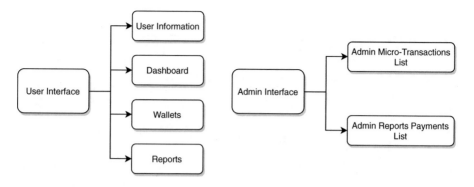

Fig. 5. User and admin interfaces and modules

4.1 User Interface

The user interface is divided in 4 main modules: user information, dashboard, wallets, and reports. In the user information module, the user can edit and verify all information inserted when signed up in the platform such as username, name, phone number, email, address, country, birth date, city and id card. The Fig. 6 presents the output of the user information module.

≡ User Information		About Us	Logout

First Name: Rui
Last Name: Carreira
Email Address: rui.33.alexandre@hotmail.com
Phone Number: 935393440
Birth Date: 12/06/2019
Country: Portugal
City: Fão
ID Card: 999999999
Address: Fão, Rua Artur Sobral nº3A
Username: bino3x

Edit   Change Password

Fig. 6. User information module

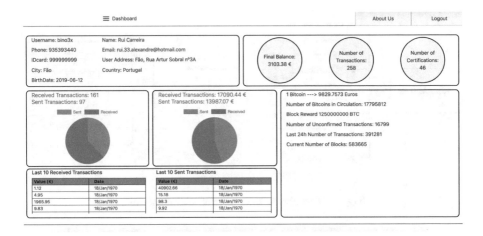

Fig. 7. Dashboard module

In the dashboard module, the output is as presented in Fig. 7. In this module the user is able to check the final balance, the number of received and sent transactions and their values, and the number of wallet ownership reports (taking in consideration all the wallets). The user is also able to verify information about the Bitcoin Network such as the bitcoin value in euros, the number of bitcoins in circulation, the reward for miners, number of unconfirmed transactions, last 24 h number of transactions and current number of blocks.

In the wallets module, the user is able to verify each wallet information, associate a new wallet and request an Wallet Ownership Report for a specific wallet. An output of this module is presented in Fig. 8.

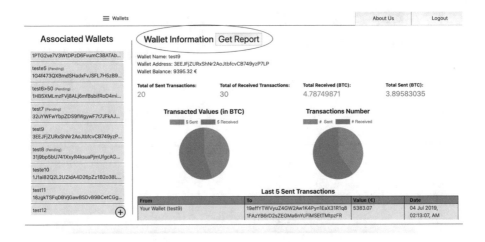

Fig. 8. Wallets module and "Get Report" option

In the wallets module, each wallet presented has two possible status: "Pending" and "done". If the status is "Pending" the user cannot verify the wallet information until he creates the micro-payment in order to prove its wallet ownership. So, the "pending" status means that the user have requested an wallet association and, in order to the wallet be associated with the user, it should give the proof of ownership through the micro transaction. If the status is done, the user has already performed the micro transaction and it is able to select the validated wallet to verify its information.

When selecting a valid wallet (in case micro-payment is already done), the user is able to check the wallet statistics and information as presented in Fig. 8, including the wallet chosen name, the wallet address, the final balance, the total number of sent and received transactions, and the total values sent and received. Also, the user is able to verify the last 5 sent and received transactions for a selected wallet, their transacted value (received or sent) and date.

Having selected one wallet from the list, the user can select the deepness of the graph and other desired characteristics, namely the number of levels and the number of transactions processed per level. This is accomplished by using a recursive algorithm that collects publicly available information from a Blockchain web-based API regarding wallets and the transactions between them. This algorithm is presented Fig. 9.

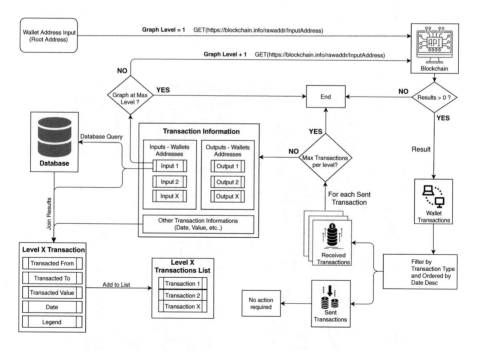

Fig. 9. Recursive algorithm diagram

The algorithm intends to collect information on different wallets, their transactions and respective relation and its operation is as follows. The algorithm begins by querying the Blockchain API on the transactions of a root wallet. In case associated transactions exist for/to the root wallet, the algorithm will filter them between sent and received transactions and then, it will order them by date. From these filtered and ordered received transactions, the algorithm picks the defined max number of received transactions defined per level. Each transaction has several inputs which correspond to the origin wallets addresses that are sending the bitcoins on the same transaction. The algorithm picks the first input of the transaction wallet addresses, identifies it as the transaction wallet address origin and verifies on the database if this transaction origin address corresponds to any wallet from any registered user on the application and its information. This database query returns if the wallet belongs to the user, if it is not recognized, or if the owner of the wallet is registered in the application and has already been issued a report or not. This status result, the previously defined transaction origin and other transaction information are merged into a transaction (node) which is presented on the graph. After finishing the first level, and with the defined maximum number of transactions/nodes per level, it is verified if the maximum number of levels has been reached. If not, the algorithm will recursively repeat the above procedures for each transaction origin of each node on the current level until it finishes.

An example for the output of this algorithm using 2 levels and 4 transactions per level is presented in the Fig. 10. After choosing the recursive graphs characteristics, the user can also request for an wallet ownership report by click on the

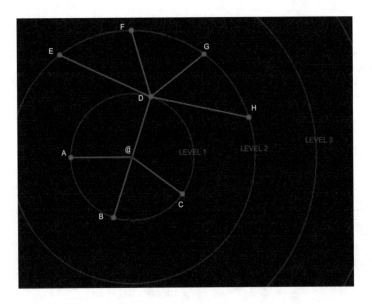

Fig. 10. Recursive algorithm output: 2 levels; 4 transactions/level

button "Get Report". The user is asked to pay the report in order for it to be generated.

The Reports module lists all the reports requested and their status, "Pending" if not yet paid, and "Issued" in case a report has already been paid and issued. The output is presented in the Fig. 11. The user can selects a Valid Wallet Ownership Report and check its contents and download it in PDF format.

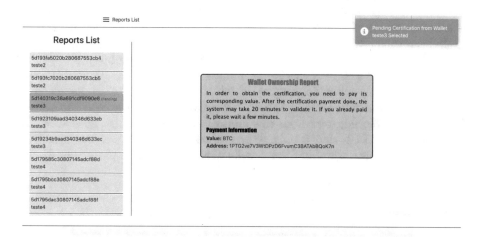

Fig. 11. Reports module

4.2   Admin Interface

The admin interface is composed of the Admin Micro-Transactions List and Admin Reports Payments List modules. In the Admin Micro-Transactions List module the administrator can check a list of all the wallets associations requests and if they are "pending" or "confirmed". This module allows also the administrator to validate the wallet micro-transactions and confirm the association requests, turning the state of the wallets from "pending" to "confirmed".

In the Admin Reports Payments List module, the administrator can check a list of all the Wallets Ownership Reports requested and if they are "pending" or "confirmed". This module allows also the administrator to validate the Wallet Ownership Report, allowing the user of the respective wallet to have access to the Wallet Ownership Report and download it.

5   Conclusions and Future Work

The cryptocurrencies such as Bitcoin are currently being used to pay products and services. In some specific scenarios, users' payments are made using cryptocurrencies and may require the user to provide a proof of ownership and proof of provenance for these payments.

In this paper a novel framework based on a web solution is proposed to generate reports upon request, intended to serve as proof of ownership and proof of provenance. Proof of ownership is accomplished using micro-payments to the respective wallets. When used recursively, the framework can produce provenance reports at request and up to a defined granularity level. This framework was tested with Bitcoin cryptocurrency and their respective transactions in euros value, but it allows also other cryptocurrencies such as Ethereum or Riple.

However, some limitations of this platform can be highlighted and future work can be addressed. The proposed framework is based on the public information available on the bitcoin network and all the information comes from an external API. There are free APIs that allow queries to bitcoin network but most of them have a very small rate of requests per minute. The Blockchain APIs in [15] are free but they only allow 600 requests each 6 min. However, if higher deep of levels and transactions is needed, paid APIs can be used or even a Blockchain node can be deployed, to prevent this limitation. Another limit imposed by the use of an open API is that it only returns a maximum of 50 transactions. Two way factor authentication and the certification of the final reports can also be addressed as future work.

References

1. Nakamoto, S.: Bitcoin: a peer-to-peer electronic cash system (2008). https://web. archive.org/web/20200224180732/https://bitcoin.org/bitcoin.pdf
2. Gemmell, P.S.: Traceable E-Cash - IEEE spectrum (1999). https://spectrum.ieee. org/computing/software/traceable-ecash
3. Fench, T., Stettner, B.: Anti-money laundering regulation of cryptocurrency: U.S. and global approaches: anti money laundering 2019 (2019). http://iclg. com/practice-areas/anti-money-laundering-laws-and-regulations/3-anti-money-laundering-regulation-of-cryptocurrency-u-s-and-global-approaches
4. Mizrahi, A.: SEC solicits blockchain analysis tool to identify wallet owners (2019). https://news.bitcoin.com/sec-solicits-blockchain-analysis-tool-to-identify-wallet-owners/
5. De La Rosa, J., Gibociv, D.: A survey of blockchain technologies for open innovation. Easychair preprint (830) (2019)
6. Gatteschi, V., Lamberti, F., Demartini, C., Pranteda, C., Santamaria, V.: To blockchain or not to blockchain: that is the question. IT Prof. **20**(2), 62–74 (2018)
7. Lin, I.C., Liao, T.C.: A survey of blockchain security issues and challenges. IJ Netw. Secur. **19**(5), 653–659 (2017)
8. Crosby, M., Pattanayak, P., Verma, S., Kalyanaraman, V.: Blockchain technology: beyond bitcoin. Appl. Innov. **2**(6–10), 71 (2016)
9. Sidhu, J.: Syscoin: a peer-to-peer electronic cash system with blockchain-based services for e-business. In: 2017 26th International Conference on Computer Communication and Networks (ICCCN), pp. 1–6 (July 2017)
10. Liang, X., Shetty, S., Tosh, D., Kamhoua, C., Kwiat, K., Njilla, L.: Provchain: a blockchain-based data provenance architecture in cloud environment with enhanced privacy and availability. In: 2017 17th IEEE/ACM International Symposium on Cluster, Cloud and Grid Computing (CCGRID), pp. 468–477 (2017)

11. Elliptic: The bitcoin big bang (2015). https://info.elliptic.co/hubfs/big-bang/bigbang-v1.html
12. Coinfirm: Coinfirm. https://www.coinfirm.com/
13. Passport.js. http://www.passportjs.org/
14. RFC 7519 - JSON Web Token (JWT). https://tools.ietf.org/html/rfc7519
15. Blockchain API: Bitcoin API - blockchain. https://www.blockchain.com/api

Transaction Costs and the Influence of New Technologies on Organizational Models

Javier Parra-Domínguez[1,2](✉), Jorge González[1](✉),
María E. Pérez-Pons[1](✉), Juan Manuel Corchado[1,2,3](✉),
and Sara Rodríguez-González[1,2,3](✉)

[1] BISITE Research Group, University of Salamanca, Salamanca, Spain
{javierparra, jorgegonzalez, eugenia.perez,
corchado, srg}@usal.es
[2] IoT Digital Innovation Hub, University of Salamanca, Salamanca, Spain
[3] AIR Institute, University of Salamanca, Salamanca, Spain

Abstract. Every company needs an organizational structure that allows it to organize and manage all its departments. For decades the most popular system of organization has been the vertical organizational model, but nowadays, thanks to the irruption of new technologies and the study of new work methodologies, other models such as the horizontal or the circular are very important. Transaction costs and the adoption of responsibilities in each of the organizational models are two key pieces in establishing the organizational model of the company. Depending on the priorities established by the entrepreneur, one organizational model or another will be chosen.

Keywords: Organizational model · Transaction costs · New technologies

1 Introduction

The type of organizational structure selected by a company plays a fundamental role in the performance of its activity [1, 19]. The choice of one type of organizational structure or another can be a differentiating element for a company to meet the established objectives [13]. Different studies show that a large number of business projects do not meet the established periods or the agreed budget, in particular, this can be observed in the study published in 1998 by "The Standish Group" where only 26% of the projects studied were developed in an adequate manner in time and costs [18].

Depending on the type of organizational structure chosen, employees may feel more or less identified with their company, and as shown by Bartel and Riketta [3, 16], employees who identify strongly with their organization may have higher job satisfaction, a higher degree of cooperation with their colleagues and a lower rate of absenteeism [4].

It is also important to assess the implications that the organization has in terms of assuming less organizational costs from the point of view of generating positive externalities [6]. Under the assumption of the Coase Theorem, the facilitation of organizational systems by new technologies will reduce the existence of legal barriers that will affect greater rigidity and costs.

J. Prieto et al. (Eds.): BLOCKCHAIN 2020, AISC 1238, pp. 144–150, 2020.
https://doi.org/10.1007/978-3-030-52535-4_15

The increased competition and complexity of today's business world makes it key to choose the right organizational structure, which allows for adequate goal setting within business projects, effective communication between different organizational levels or adequate control over the degree of project development.

In this article we focus on the study of vertical and horizontal organization models and their relationship with new technologies, applying an economic perspective to these organizational models, previously carrying out a review of the different organizational models that exist. The article is structured as follows; in point two we introduce the different existing organizational models and in point three we work on the economic meaning and influence of the new technologies on the different organizational models. Finally, we establish the conclusions.

2 Organizational Models

There are numerous organizational models within a company and organization charts are commonly used to represent them [5]. Organizational charts are essentially the graphic representation of a company and reflect in a schematic way the areas or departments it is made up of and their hierarchy levels. If we consider authors such as Elio Rafael de Zuani or Enrique Benjamin Franklin, we can distinguish organization charts in five different categories: according to their nature, according to their purpose, according to their scope, according to their content and according to their presentation. As a reference for the development of this section we have chosen those implemented by Fincowsky and Benjamin [11].

In this article we focus on vertical organizational models, so we study the organization charts according to their presentation [8]. In this category of organization charts we can find five different configurations:

- Vertical organization chart: in this type of organization chart, director is placed on the top step and the hierarchical levels are disaggregated from top to bottom. It is the most common organizational model within companies [7].
- Horizontal organization chart: In this case, the headline is placed on the left side of the organization chart and the hierarchical levels are broken down in columns from left to right [10].
- Mixed organization chart: this is a type of organization chart that uses a combination of vertical and horizontal organization charts and is suitable for companies that have many elements in the base [10].
- Block organization chart: type of organization chart that constitutes a variant of the vertical organization chart and is characterized by integrating more units in smaller spaces [10].
- Circular organization chart: in this type of flowchart the hierarchy extends from the centre of the circle to the outside in concentric circles. Each of the circles represents a level of authority and the units of equal hierarchy are located in the same circle [10] (Figs. 1, 2 and 3).

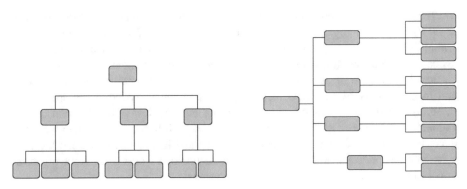

Fig. 1. Vertical organization chart   Fig. 2. Horizontal organization chart

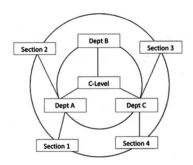

Fig. 3. Circular organization chart

As mentioned above, the vertical organization chart is the most commonly used in companies as it represents a hierarchical-based business relationship and is based on a pyramid-type structure. The main characteristics of this type of organigram are:

- The company's management and production lines are clearly differentiated.
- Appearance of control by showing a clear organization of the company.
- Motivates the team to move up in the company.
- Workers at the base of the organization could feel less valued.
- Decision-making is delayed until it reaches the decision-making level.
- The leader at the top of the organization must know how to manage the company to lead it to success.

In addition, the horizontal organization chart has the following characteristics:

- Employees have the ability to make decisions without having to wait for explicit orders from their superiors.
- There is a high level of fluidity in collaboration between employees of the same department and with other departments.
- There is a high level of fluidity in communication between employees.
- The teams have greater responsibility.

- Leaders have less authority.
- There may be a lack of accountability affecting the whole team if the work does not go as expected.

3 Economic Sense and the Influence of New Technologies on Organizational Models

3.1 Transaction Costs and Organizational Models

In the previous section we have described the two main organizational models according to their organization, the vertical and the horizontal. These two models have different characteristics and one of the most important differences is the way in which decisions are made. While in the vertical organizational model a constant transmission of information to a superior is necessary in order to execute decisions, in the horizontal organizational model there is autonomy of decisions between the different teams that form part of the company.

We analyze this phenomenon from the point of view of the theory of transaction costs. Coase's theorem states that if property rights are well defined and transaction costs are zero, trading between parties will lead to an optimal market allocation point [12]. If we apply this theorem to organizational models we could see that in the vertical organizational model there are clearly transaction costs, since it is essential to constantly communicate information to superiors for decision-making and therefore it is possible that the desired results are not being achieved in the company due to the costs involved in decision-making. In this case an optimal allocation of resources would not be taking place and the Coase theorem would not be fulfilled.

However, in horizontal organizational models there is no superior to the one who needs to communicate information constantly in order to make decisions. Instead, teams make decisions autonomously, which would eliminate these transaction costs, increase the efficiency of the company and optimally allocate resources.

The assumptions of the Coase Theorem are twofold, as mentioned above, that the costs of any negotiation are low for both parties, and that the owner of the resources can identify the consequence of using them [7]. This is why, in addition to the fact that autonomous decision systems have an impact on the elimination of transaction costs, which largely favours the appearance of more decentralised horizontal organisations, we can theoretically state that the incorporation of new technologies favours an optimisation in the location of the company's resources. It is also important to note that, from a market point of view, Coase expressed public intervention as necessary when there were no established property rights. In this sense, the new technologies make it easier to have a clear understanding of the responsibility between departments and to be able to direct it towards one or other person, thus avoiding conflicts that imply the intervention of general management in problems understood as externalities [2, 9].

3.2   Organizational Models and New Technologies

With the emergence of new technologies, the way in which people communicate within companies has been simplified, making communication more convenient and efficient. In addition, they facilitate the adoption of skills and experience needed to develop the tasks required by the company. These technological advances also have effects on the organizational models we are studying.

If we consider the vertical model of organization, the emergence of new technologies can simplify the process of communication of information between different levels of hierarchy for decision making, since the person who occupies a higher level of hierarchy has more facility to collect information from lower levels in less time, which would speed up decision-making and allow a reduction in transaction costs, in addition to limiting the loss of information that usually occurs when transmitting it between different hierarchical levels.

In recent years, when information technologies and the globalization of the digital world have developed with greater intensity, interest in horizontal organizations has grown as large companies in the world of technology and distribution have chosen to adopt this type of organization. In these new organizations, a decentralized organizational model is being developed, with little hierarchy and where teamwork and information technologies predominate [14].

As in vertical organizations, new technologies allow for a better adoption of skills and experience in horizontal organizations, and allow for a new perspective in the search for workers for companies; it is not only a matter of looking for a person to fill the vacant job position, but also of having the right people within the organization occupy the right jobs according to their skills and abilities [15].

With the new ICTs, decision making in horizontal organizations continues to be decentralized but the elimination of hierarchical levels can concentrate knowledge and decision making in certain positions. Because of this, training of workers and supervision of tasks is essential, something that is simplified by the new technologies and allows workers to obtain a greater degree of knowledge and responsibility [13].

Finally, new technologies affect a key aspect of organizations, communication. New technologies allow direct contact with teammates, suppliers, or business partners, and this makes the transmission of information faster and more effective.

4   Conclusions

The two organizational models studied in the previous sections imply significant differences in decision making. While vertical organizational models have higher transaction costs and have a hierarchical structure in decision-making, horizontal organizational models have lower transaction costs and teams can make decisions autonomously.

If we only consider the transaction costs, it would be preferable to choose a horizontal organization model as it will allow faster decision making and more effective communication. However, we could also take into account the responsibilities that exist in companies.

In a company with a vertical organization model, the superior in the hierarchy is always responsible for making decisions and if the project objectives are not met, the responsibility will fall on him. On the other hand, in horizontal organizational models there may be a lack of accountability on the part of the entire team in case the project results are not what is expected [17]. Because of this, in a vertical organizational model there may be a greater sense of accountability in decision making than in a horizontal organizational model.

According to the trends observed in recent years, it is to be expected that a large part of future organizations will adopt a horizontal structure, since there is a growing trend towards the configuration of equipment within companies and the use of digital media to develop work activity, which favors this type of organization.

5 Discussions

Discussions
Although the study focuses on the positive impact that technology can have on the development of organizations from the point of view of horizontal or vertical structures, the impact can be considered negative if, empirically, it is shown that it does not adapt well to the organizational model presented.

From the point of view of the implementation of agile methodologies, technology supports and enriches organizations, whether they are vertical or horizontal, but we must continuously think about the development of people through technology, and it is in the perfect coupling of technology and human development where we will truly observe if technology couples and improves the vertical structures.

Acknowledgements. This work has been partially supported by the European Regional Development Fund (ERDF) through the Interreg Spain-Portugal V-A Program (POCTEP) under grant 0677_DISRUPTIVE_2_E (Intensifying the activity of Digital Innovation Hubs within the PocTep region to boost the development of disruptive and last generation ICTs through cross-border cooperation).

References

1. Ashton, D.N.: The impact of organisational structure and practices on learning in the workplace. Int. J. Train. Dev. **8**(1), 43–53 (2004)
2. Azariadis, C., Drazen, A.: Threshold externalities in economic development. Q. J. Econ. **105**(2), 501–526 (1990)
3. Bartel, C.A.: Social comparisons in boundary-spanning work: effects of community outreach on members' organizational identity and identification. Adm. Sci. Q. **46**(3), 379–413 (2001)
4. Bartels, J., Peters, O., de Jong, M., Pruyn, A., van der Molen, M.: Horizontal and vertical communication as determinants of professional and organisational identification. Pers. Rev. **39**(2), 210–226 (2010)
5. Cassia, L., Paleari, S., Redondi, R.: Management accounting systems and organisational structure. Small Bus. Econ. **25**(4), 373–391 (2005)
6. Cooter, R.: The cost of coase. J. Leg. Stud. **11**(1), 1–33 (1982)

7. Cooter, R.D.: The coase theorem. In: Allocation, Information and Markets, pp. 64–70. Palgrave Macmillan, London (1989)
8. Colombo, M.G., Croce, A., Murtinu, S.: Ownership structure, horizontal agency costs and the performance of high-tech entrepreneurial firms. Small Bus. Econ. **42**(2), 265–282 (2014)
9. Dalkir, K.: Knowledge Management in Theory and Practice. Routledge, Abingdon (2013)
10. De Zuani Elio, R.: Introducción a la Administración de Organizaciones. Editorial Maktub, Salta (2003)
11. Fincowsky, F., Benjamín, E.: Organización de empresas. Diseño, y Estructura, Editores McGRAW–HILL INTERAMERICANA SA DE CV (2009)
12. Gil Estallo, M.A., Griful Miquela, C., De Val Pardo, I.: From horizontal organization to knowledge integration. In: 51th International Atlantic Economic Conference, Atenas, Marzo (2001)
13. Maduenyi, S., Oke, A.O., Fadeyi, O., Ajagbe, A.M.: Impact of organisational structure on organisational performance (2015)
14. Medema, S.G.: Coase Theorem. Wiley Encyclopedia of Management, pp. 1–12 (2015)
15. Pardo, I.D.V., Estallo, Á.G.: Organización vertical versus horizontal. ESIC MARKET (2004)
16. Riketta, M.: Organizational identification: a meta-analysis. J. Vocat. Behav. **66**(2), 358–384 (2005)
17. Rubenstein-Montano, B., Liebowitz, J., Buchwalter, J., McCaw, D., Newman, B., Rebeck, K., Team, T.K.M.M.: A systems thinking framework for knowledge management. Decis. Support Syst. **31**(1), 5–16 (2001)
18. Stare, A.: The impact of the organisational structure and project organisational culture on project performance in Slovenian enterprises. Manag.: J. Contemp. Manag. Issues **16**(2), 1–22 (2011)
19. Zheng, W., Yang, B., McLean, G.N.: Linking organizational culture, structure, strategy, and organizational effectiveness: mediating role of knowledge management. J. Bus. Res. **63**(7), 763–771 (2010)

BLOCKCHAIN-DC

Analysis of Costs for Smart Contract Execution

Felix Adler[(⊠)], Dennis Kitzmann, and Marc Jansen

Computer Science Institute, University of Applied Sciences Ruhr West,
46236 Bottrop, Germany
{felix.adler,dennis.kitzmann}@stud.hs-ruhrwest.de

Abstract. This work analyses the differences in Gas cost consumption for the execution of individual functions between the two Smart Contract (SC) programming languages Solidity and Vyper. For this purpose, SCs for the most basic types *ERC-20*, *MultiSig* and *Atomic Swap* were created for both programming languages. To enable a comparison, it was ensured that a SC fulfills exactly the same task in each programming language. It turns out, that the calculated estimated Gas amount from the compiler is mostly meaningless. There is also no programming language, that is always superior to the other in Gas cost efficiency for each SC type and that Waves can be an alternative.

Keywords: Gas cost · Smart Contracts · Ethereum · Waves · Blockchain

1 Introduction and Related Work

Blockchain [1] based systems have gained popularity in recent years. The Ethereum [2] and Waves blockchain, which support SCs [3], has emerged as an interesting development. SCs, whose concept was introduced by Nick Szabo, are programs that can be executed on a blockchain. These programs can automatically trigger, control and document legally relevant actions depending on digitally verifiable events. *ERC-20*, *MultiSig*[1] and *Atomic Swap* [4] have established themselves as some of the basic SC types, which can be created on the Ethereum blockchain in one of the available programming languages Solidity[2] and Vyper[3]. For Waves, SCs can be created using the functional programming language RIDE[4]. As described in a similar paper *The Economics of Smart Contracts* [5], all transactions on Ethereum are accompanied by commission payments, the amount of which depends on the current load on the network. This fee is paid in value units called *Gas*. Since several programming languages are available on

---

[1] https://bitcoinmagazine.com/articles/multisig-future-bitcoin-1394686504.
[2] https://solidity.readthedocs.io/.
[3] https://vyper.readthedocs.io/.
[4] https://wavesplatform.com/files/docs/white_paper_waves_smart_contracts.pdf.

© The Editor(s) (if applicable) and The Author(s), under exclusive license to Springer Nature Switzerland AG 2020
J. Prieto et al. (Eds.): BLOCKCHAIN 2020, AISC 1238, pp. 153–156, 2020.
https://doi.org/10.1007/978-3-030-52535-4_16

Ethereum and each execution of a SC costs Gas, it is advantageous to select the most economical programming language for the respective SC type. On Waves there is no Gas, but each transaction with script execution, costs basically 0.005 Waves[5].

2   Research Question and Analysis Strategy

The research question that drove the motivation of the presented analysis, was to determine the Gas costs for the execution of individual functions of the SC programming languages Solidity and Vyper for the most basic types *ERC-20*, *MultiSig* and *Atomic Swap* on the Ethereum blockchain, to find out, which programming language is generally more economical for the respective type. A related aspect, raised by this research question, is whether Waves can be an economic alternative to Ethereum. In the long term, it is advantageous to keep Gas costs as low as possible when executing a SC. That is why the right choice of the programming language is of particular importance. SCs of each type in Solidity and Vyper were created for evaluation[6]. To enable a comparison, it was ensured that a SC fulfills exactly the same task in each programming language. The *ERC-20 Basic* SC corresponds to the ERC-20 Token Standard[7]. With the *MultiSig Basic* SC, two addresses must sign to verify a proof. Two variants were created for the *Atomic Swap*. One allows Atomic Swaps on different blockchains that use Ether. The other allows Atomic Swaps between ERC-20 SCs. For the estimated Gas amount, the compiler output or the ABI was used, when the SCs were translated into bytecode. The real Gas amount were determined by transactions with the same test data sets for Solidity and Vyper. Testing was done on 2020-02-07 within the Remix IDE using the JavaScript VM test environment. The Solidity compiler version used was `0.6.1+commit.e6f7d5a4`. For Vyper, the remote compiler version was used on the test day within the Remix IDE. When different Gas costs are incurred in different programming languages for the same functionality, a statement can be made about the economic efficiency of the programming language in the current context. To evaluate Waves' economy, a cost comparison with Ethereum was made using history data from 2020-01-01 and the results converted to Gas on that day. In addition, the unit price *Break Even Point* (BEP) of Waves was calculated with the same history data for each SC function from Solidity's and Vyper's average real gas costs. The BEP indicates the unit price for Waves in dollars, from which the two blockchains have approximately the same cost.

3   Evaluation Results

Table 1 shows the Gas amount for the execution of each function of a SC for Solidity and Vyper and Waves' BEP. As it turns out, the estimated Gas amount

---

[5] https://docs.wavesplatform.com/en/blockchain/transaction/transaction-fee# smart-accounts.

[6] https://gitlab.hs-ruhrwest.de/fyfeadle/blockchain-2020-doctoral-consortium.

[7] https://eips.ethereum.org/EIPS/eip-20.

is basically calculated incorrectly. For example, in the `transfer` function, the percentage difference between the real and estimated Gas amount for Vyper is 937.43%. At *Atomic Swap ERC-20 to ERC-20*, transactions are executed across several connected ERC-20 SCs. Here, Solidity is in total 20.83% more economical than Vyper. At *Atomic Swap Ether*, the Solidity compiler cannot predict a single value. At this SC, Vyper is in every function a little and in total 3.54% more economical than Solidity. At *MultiSig Basic*, the cost difference between the two programming languages is not too big. This could be due to the fact, that this SC is not particularly complex compared to the other types. The *ERC-20 Basic* shows, that the Solidity compiler has correctly precalculated the Gas amount for the getter functions `allowance`, `balanceOf` and `totalSupply`. The Vyper compiler, on the other hand, predicts a too low value for the mentioned functions. Otherwise, the cost difference is really small. The average BEP unit price of Waves for each SC is always higher than the current unit price. For example for the function `allowance`, the average of real Gas amounts is 1.276 Gas. To get this amount of Gas, Waves unit price should be $0.389 (see Table 2 for further information) → The costs would be about the same for both blockchains. Since tokens like ERC-20 are implemented natively in Waves and do not have to be created via a SC, the costs are even lower.

Table 1. Result matrix of the Gas amount analysis with Waves' BEP unit price.

Type	Function	Solidity est. Gas	Solidity real Gas	Vyper est. Gas	Vyper real Gas	Waves BEP
ERC-20 Basic	allowance	1,409	1,409	821	1,143	$0.389
	approve	22,384	3,184	37,822	2,963	$0.938
	balanceOf	1,240	1,240	817	1,188	$0.370
	totalSupply	1,013	1,013	573	995	$0.306
	transfer	infinite	6,996	74,207	7,916	$2.275
	transferFrom	infinite	10,076	110,074	10,564	$3.149
	Gas sum	**infinite**	**23,918**	**224,314**	**24,769**	**ø $1.237**
MultiSig Basic	sign	21,383	22,925	36,291	22,711	$6.963
	verify	42,613	17,413	42,097	15,259	$4.985
	Gas sum	**63,996**	**40,338**	**78,388**	**37,970**	**ø $5.974**
Atomic Swap Ether	open	infinite	107,746	248,935	104,727	$32.418
	close	infinite	37,067	111,275	35,451	$11.064
	expire	infinite	15,350	74,590	14,255	$4.517
	check	infinite	5,793	2,462	5,621	$1.741
	checkSecretKey	infinite	6,707	2,856	6,501	$2.015
	Gas sum	**infinite**	**172,663**	**440,118**	**166,555**	**ø $10.35**
Atomic Swap ERC-20 to ERC-20	open	infinite	124,244	287,799	144,176	$40.954
	close	infinite	37,386	116,163	64,895	$15.605
	expire	infinite	22,454	76,815	25,271	$7.282
	check	2,323	5,923	2,480	5,642	$1.765
	Gas sum	**infinite**	**190,007**	**483,257**	**239,984**	**ø $16.40**

As already mentioned in the introduction, the cost of a Waves transaction with script execution is basically 0.005 Waves. A transaction is only possible as long as the maximum script complexity[8] of 4000 is not exceeded. Table 2 shows the cost of a Waves transaction in dollars and the Gas, available for that amount. All data were collected on 2020-01-01 and the average Gas price on that day was

---

[8] https://docs.wavesplatform.com/en/ride/base-concepts/complexity.

11.664 Gwei[9]. However, the value of Waves plays a major role. If the marketcap of Waves is notionally raised to that of Ethereum, transaction costs would be very expensive.

Table 2. Cost comparison table between Ethereum and Waves from 2020-01-01.

Blockchain	Marketcap	Circulating supply	Price per unit	Waves transact. price
Ethereum	$14,271,059,633	109,104,324	$130.802	$0.005 (3,435 Gas)
Waves	$105,609,699	100,754,836	$1.05	
Ethereum	$14,271,059,633	109,104,324	$130.802	$0.708 (464,191 Gas)
Waves (fictive)	$14,271,059,633	100,754,836	$141.641	

4   Reflections

As the analysis shows, there is no programming language, that is always superior to the other in Gas cost efficiency for each SC type and that Waves can be an alternative. The analysis also raises the open question of how code optimization [6] and *Clean Code* [7] compliance affect gas cost efficiency. This problem is not relevant for Waves, since costs are basically constant.

References

1. Nakamoto, S.: Bitcoin: a peer-to-peer electronic cash system (2008). https://bitcoin.org/bitcoin.pdf. Accessed 07 Feb 2020
2. Buterin, V.: A next generation smart contract and decentralized application platform (2014). https://cryptorating.eu/whitepapers/Ethereum/Ethereum_white_paper.pdf. Accessed 07 Feb 2020
3. Szabo, N.: The idea of smart contracts (1997). https://nakamotoinstitute.org/the-idea-of-smart-contracts. Accessed 07 Feb 2020
4. Mahdi, H., Miraz, D.C.D.: Atomic cross-chain swaps: development, trajectory and potential of non-monetary digital token swap facilities (2019). https://arxiv.org/abs/1902.04471. Accessed 07 Feb 2020
5. Baird, K., Jeong, S., Kim, Y., Burgstaller, B., Scholz, B.: The economics of smart contracts (2019). https://arxiv.org/abs/1910.11143. Accessed 07 Feb 2020
6. Chen, T., Li, X., Luo, X., Zhang, X.: Under-optimized smart contracts devour your money (2017). https://arxiv.org/abs/1703.03994. Accessed 07 Feb 2020
7. Martin, R.C.: Clean Code: A Handbook of Agile Software Craftsmanship. Prentice Hall, Upper Saddle River (2009)

---

[9] General coin data: https://coinmarketcap.com/
Ethereum related: https://etherscan.io/.

Privacy in Financial Information Networks: Directions for the Development of Legal Privacy-Enhancing Financial Technologies

Valeria Ferrari[✉]

University of Amsterdam, Nieuwe Achtergracht 166,
1018 WV Amsterdam, The Netherlands
v.ferrari@uva.nl

Abstract. In light of the strategic role of financial information for law enforcement, the protection of privacy regarding financial data must be balanced with the advantages of automated mechanisms for the monitoring and recording of financial activities. The growing availability of financial data and the global dimension of financial networks, however, impose to carefully examine practices concerning the management of financial data, checking them against the core rules and principles of data protection. This paper steers this effort providing a framework for understanding privacy in the financial context; moreover, it overviews the relevant legal frameworks affecting the management of financial information and assesses concrete industry practices to expose some compelling privacy issues. The study suggests that further research is needed to establish (1) whether current practices in the financial industry determine a lack of legal protection with regard to users' privacy; (2) whether it is desirable to develop technological solutions that allow a greater degree of anonymity for digital financial transactions; and, if so, (3) which is the most suitable governance/legal and institutional framework for anonymous digital transactions.

Keywords: Financial privacy · Cryptocurrencies · FinTech · GDPR

1 Introduction

In the past decades, two major tendencies have emerged that urge to bring the issue on financial privacy in the spotlight. The first one is the digitalization of money and commerce, which have exponentially expanded the production and availability of financial data. In 2019, countries like Sweden and the Netherlands have registered a higher total amount of digital transactions than cash-based ones, showing a tendency towards substituting cash even in daily small-size payments. The second tendency is the reconfiguration of the incentives underlying the provision of financial services around data exploitation.

New tools for data collection and processing and possibilities of intersecting financial data with additional information about users' online activities situate financial informational within the logics of contemporary information economy. Financial intermediaries are subject to sector-specific provisions that – pursuing the objectives of

J. Prieto et al. (Eds.): BLOCKCHAIN 2020, AISC 1238, pp. 157–160, 2020.
https://doi.org/10.1007/978-3-030-52535-4_17

transparency, prevention of illegal activities and tax avoidance – mandate data collection, data retention and reporting obligations regarding individuals' transactions in order to enable efficient enforcement. These rules and the goals they pursue seem to overrun privacy considerations; however, the present paper argues that ongoing development of a the financial industry – characterized by data-intensive business models and global reach of financial intermediaries - impose to re-define the meaning of privacy in this particular context, checking the practices and the policies that govern financial information against the core principles of data protection.

2 Methodology

This paper addresses issues of financial privacy scrutinizing (a) the incentives for information gathering and the power imbalances created by data flows; and (b) the transparency and fairness of automated data processing and algorithm-based decision-making. The study provides an overview of legal instruments that affect the governance of financial data: Anti-Money Laundering (AML) and Counter-Terrorist Financing (CTF) policies on one side, and privacy legislation - i.e. the GDPR[1] and the Law Enforcement Directive[2] - on the other. In this context, the technological means that are or can be deployed to achieve the legal objectives are presented. Finally, the study discusses practices in the management of financial data that raise privacy issues and/or expose conflicts between law enforcement priorities and privacy legal protections.

3 Preliminary Findings

3.1 (Conflicting) Legal Frameworks Governing the Flows of Financial Data

Regulatory and governance frameworks at the international and European level ensure that financial institutions and firms cooperate with law enforcement agencies providing access to financial databases. The coordinating guidelines of the FAFT (Financial Action Task Force) and the Common Reporting Standards (CRS) by the OECD (Organization for Economic Co-operation and Development) set out global standards for the collection and exchange of financial information in the fight against money laundering, tax evasion and other financial crimes. At the EU level, the 5th Anti-Money Laundering Directive[3] the and other legal instruments[4] mandate that financial intermediaries have in place automated systems for customer identification, transactions monitoring and reporting. These compliance processes imply massive data collection, long data retention periods and the use of algorithmic decision-making systems for consumers' profiling and red flagging. While the regime of surveillance over financial

---

[1] Regulation (EU) 2016/679 (General Data Protection Regulation).

[2] Directive (EU) 2016/680.

[3] Directive (EU) 2015/849.

[4] E.g. Directive (EU) 2015/2366, Directive 2006/24/EC, etc.

flows has been strengthen in the aftermath of the 2008 financial crisis, the adoption of the GDPR has, in the EU, introduced principles and priorities for the government of business data that are diametrical opposite to those set out by AML and CTF rules. Financial institutions, therefore, are expected to enforce legal requirements and policy goals that are uneasy to incorporate within the same technological and governance structure.

In the meantime, solutions based on distributed ledger technologies have been proposed to address both the problems of financial transparency and privacy, following a different logic: not by enhancing the responsibility of the intermediary to enforce the legal objective, but eliminating vulnerabilities in the system by decentralizing its governance.

3.2   Privacy Issues in Financial Information Networks

Both the GDPR and the Police Directive create exceptions to privacy protection regimes when data is collected and analyzed for crime prevention, investigation and other legitimate law enforcement necessities. However, the boundaries of such limitations are not well defined: industry and law enforcement practices must be continually checked against the principles and standards set out by the European privacy frameworks. The Article 29 Data Protection Working Party (WP29) stresses, in various documents, that the core principles of data protection - including the principle of purpose limitation and data minimization - must be applied to data acquired and stored for law enforcement purposes.[5] This is not an easy task: the global dimension and the technological development of financial information networks make the EU legal instruments insufficient to prevent questionable practices of both financial firms and pubic authorities that affect European citizens' data.

Problems arise from the dual purpose of the personal data collected by financial intermediaries in the context of their AML/CTF procedures. Financial institutions collect and process massive consumers data on the basis of their legal obligations to do so, but it's hard to prevent such data to be used for consumer purposes as well - e.g. to provide personalized services, for targeted advertainment and credit scoring. For the latter, they share data with third parties, including insurance companies, marketing firms and social media platforms.[6] The dual purpose of the processing hampers the enforceability of individuals' rights as granted by the GDPR. The rights to data portability and erasure established by the GDPR for data collected for commercial purposes, for example, are not granted in case of data collected in the context of AML and law enforcement procedures.

Another, interrelated, aspect that affects the enforcement of European privacy policies is the cross-national nature of financial services and the underlying data flows.

---

[5] See, for instance, Article 29 Data Protection Working Party, "Opinion on some key issues of the Law Enforcement Directive (EU 2016/680), available at: https://ec.europa.eu/newsroom/article29/item-detail.cfm?item_id=610178.

[6] This is stated in some banks' and payment providers' privacy policies; some banks, for instance, share data with Facebook for the service of Facebook Custom Audience, used for targeted marketing.

Data protection rules established for firms and public authorities in the EU do not always have equivalents in the US. A staggering difference is, for instance, the data retention limitation: US companies can retain consumers data for up to 80 years. Moreover, while in the EU firms are bound by the principle of purpose limitation, in the US the commercial use of data collected for enforcement purposes is not prohibited. The impact of these differences in terms of privacy, surveillance and geopolitical power imbalances becomes glaring if one considers the global pervasiveness of the US financial service industry.

Finally, the high volume of data processing involved in AML procedures obliges financial firms to deploy automated or semi-automated systems for data collection and analysis and algorithm-based consumers profiling. Again, these practices are specifically addressed by the GDPR, but are not comprehensively targeted by US law. The WP29 has underlined how profiling, even when deployed in the context of law enforcement activities, must respect data protection principles and be grounded on a legal basis specified by national law.[7] Moreover, while the GDPR qualifies AML as legitimate grounds for automatic processing, the regulation guarantees nonetheless the individual's right to challenge the outcome of such processing. Once again, however, the expansion of data availability (including the possibility to link financial data with other types of personal data), the increasing complexity of algorithmic systems and and the global dimension of information networks challenge the effectiveness of European legal safeguards.

References

1. Campbell-Verduyn, M.: Bitcoin, crypto-coins, and global anti-money laundering governance. Crime, Law and Social Change, vol. 69 (2018)
2. Chaum, D.: Achieving Electronic Privacy. Scientific American, New York (1992)
3. Cohen, J.E.: What Privacy Is For. Harvard Law Review, vol. 126 (2013)
4. Frasher, M.: Multinational banking and conflicts among us-euaml/ctf compliance &privacy law: operational & political views in context. Swift Institute Working Paper No. 2014-008 (2016)
5. Gurses, S., Hoboken, J.: Privacy After the Agile Turn. Cambridge Handbook of Consumer Privacy, Cambridge (2018)
6. Jentzsch, N.: Financila privacy: An International Comparison of Credit Reporting Systems. Springer, Heidelberg (2007)
7. Lloyd, I.: Privacy, anonymity and the Internet. Electron. J. Comparat. Law 13, 1 (2009)
8. Stan, S.: Financial Privacy in a Cashless Society (2019). SSRN: https://ssrn.com/abstract=3367610 or http://dx.doi.org/10.2139/ssrn.3367610
9. Swire, P.P.: Financial Privacy and the Theory of High-Tech Government Surveillance, Washington University Law Review, vol. 77 (1999)
10. Kleiman, M.N.: Privacy vs computerised Law Enforcement, Northwestern University Law Review, vol. 86 (1992). http://www.springer.com/lncs. Accessed 21 Nov 2016

---

[7] Article 29 Data Protection Working Party, "Guidelines on Automated Individual Decision-Making and Profiling for the Purposes of Regulation 2016/679", available at: https://ec.europa.eu/newsroom/article29/item-detail.cfm?item_id=612053.

Energy Markets with Blockchain Technology

Yeray Mezquita[(✉)] [ID]

BISITE Research Group, University of Salamanca, Salamanca, Spain
yeraymm@usal.es

Abstract. The ease of access to renewable energy production by users has made possible the emergence of individuals who are self-sufficient and sell their energy to the grid, prosumers. This fact is implying a change in the current paradigm, in which energy can be acquired directly from individuals and not necessarily from the multinationals that monopolise the market. However, so that this change can be carried out, solving the risks that any type of system connected to Internet has, it is possible to make use of the blockchain technology, which it is needed more research to being mass adopted.

Keywords: Blockchain · Energy markets · Multi-agent systems · Security

1 Introduction

Although electricity consumption is expected to increase by a third in 2035, there is currently a change in trend in the type of energy produced, reaching 25% production by means of renewable energies [20,26,32]. This upturn in renewable energy production makes the term prosumer the order of the day. The so-called prosumers are entities that produce their own energy, thanks to the increasingly easy access to renewable energies, and end up dumping the excess they do not need into the network [2,5,8,19,28]. This tendency helps to look for a greater benefit in the purchase and sale of energy on the part of the prosumers and the consumers, obtaining the first one more money by the sale of energy directly to consumers; and the second one a cheaper energy when acquiring it of the prosumers directly; all this without the electrical company as an intermediary, using only local energy microgrids [4,6,7,12,12,17,21].

The possibility of obtaining better benefits automatically based on supply and demand has been widely studied in the literature [3,10,18,29]. Most of the proposed solutions make use of intelligent agents that negotiate the price of energy. This kind of approaches have been used in other fields too with a good performance [9,13,14,16,27].

However, the use of intelligent agents that automate the trading process brings with it some risks, such as the possibility of attacks that prevent the correct exchange of information; that the information exchanged has been modified

J. Prieto et al. (Eds.): BLOCKCHAIN 2020, AISC 1238, pp. 161–164, 2020.
https://doi.org/10.1007/978-3-030-52535-4_18

during its transmission; etc. For this reason, there are works that have decided to incorporate blockchain technology in multi-agent platforms, thanks to which communication channels are secured [24].

With the use of blockchain technology it is implemented a distributed ledger on the platform that underlies [30,31], that is used by the agents as a bulletin board to publish all the information relevant to the proper functioning of the platform. Thanks to the use of smart contracts [25], it is possible the virtualization of real assets along its traceability along the process of buying and selling [1,15,23].

While the application of this technology may solve the major problems associated with the use of smart agents for the automation of energy markets, further research is needed on how to solve other challenges associated with the use of blockchain technology when implemented in sensor-based systems, like the type of blockchain technology used in the platform based on the necessities and how to tackle the attacks that might suffer the network of nodes that sustain the blockchain [11,22].

2   Conclusions

This work has studied the current trend on the creation of automatic energy markets in the field of electric microgrids. The use of multi-agent systems has spread for the development of these markets. In addition, to solve some of the problems that intelligent agents face, such as trust in the communications channel and in the messages received, the use of blockchain technology has been studied in the literature. However, this technology is not without risk and must be taken into account when using it.

As future work, it is necessary to study the viability of this type of ecosystem on a large scale, since, although in local environments it has been demonstrated that they can be implemented using public blockchain infrastructures [23], not if the world was filled with this kind of platform.

Acknowledgements. The research of Yeray Mezquita is supported by a pre-doctoral fellowship from the University of Salamanca and Banco Santander. Also, this work has been partially supported by the Salamanca Ciudad de Cultura y Saberes Foundation under the Talent Attraction Program (CHROMOSOME project).

References

1. Alvarado-Pérez, J.C., Peluffo-Ordónez, D.H., Therón-Sánchez, R., et al.: Bridging the gap between human knowledge and machine learning (2015)
2. Brondino, M., Dodero, G., Gennari, R., Melonio, A., Raccanello, D., Torello, S.: Achievement emotions and peer acceptance get together in game design at school. ADCAIJ: Adv. Distrib. Comput. Artif. Intell. J. **3**(4), 1–12 (2014)
3. Carvalhal, C., Deusdado, S., Deusdado, L.: Crawling pubmed with web agents for literature search and alerting services. ADCAIJ: Adv. Distrib. Comput. Artif. Intell. J. **1**, 19–22 (2013)

4. Casado-Vara, R., Chamoso, P., De la Prieta, F., Prieto, J., Corchado, J.M.: Non-linear adaptive closed-loop control system for improved efficiency in iot-blockchain management. Inf. Fusion **49**, 227–239 (2019)
5. Casado-Vara, R., Novais, P., Gil, A.B., Prieto, J., Corchado, J.M.: Distributed continuous-time fault estimation control for multiple devices in iot networks. IEEE Access **7**, 11972–11984 (2019)
6. Chamoso, P., González-Briones, A., Rodríguez, S., Corchado, J.M.: Tendencies of technologies and platforms in smart cities: a state-of-the-art review. Wireless Commun. Mob. Comput. **2018** (2018)
7. Choon, Y.W., Mohamad, M.S., Safaai Deris, R.M., Illias, C.K.C., Chai, L.E., Omatu, S., Corchado, J.M.: Differential bees flux balance analysis with optknock for in silico microbial strains optimization. PLoS ONE **9**(7), e102744 (2014)
8. Cofini, V., de la Prieta, F., Di Mascio, T., Gennari, R., Vittorini, P.: Design smart games with requirements, generate them with a click, and revise them with a guis. ADCAIJ: Adv. Distrib. Comput. Artif. Intell. J. **1**(3), 55–68 (2012)
9. Francisco, M., Mezquita, Y., Revollar, S., Vega, P., De Paz, J.F.: Multi-agent distributed model predictive control with fuzzy negotiation. Expert Syst. Appl. **129**, 68–83 (2019)
10. Frikha, M., Mhiri, M., Gargouri, F., et al.: A semantic social recommender system using ontologies based approach for Tunisian tourism (2015)
11. Fyfe, C., Corchado, J.M.: Automating the construction of CBR systems using kernel methods. Int. J. Intell. Syst. **16**(4), 571–586 (2001)
12. Glez-Bedia, M., Corchado, J., Corchado, E., Fyfe, C.: Analytical model for constructing deliberative agents. Eng. Intell. Syst. Electr. Eng. Commun. **10**(3), 173–185 (2002)
13. González-Briones, A., Castellanos-Garzón, J.A., Mezquita Martín, Y., Prieto, J., Corchado, J.M.: A framework for knowledge discovery from wireless sensor networks in rural environments: a crop irrigation systems case study. In: Wireless Communications and Mobile Computing 2018 (2018)
14. Guivarch, V., Camps, V., Péninou, A.: Amadeus: an adaptive multi-agent system to learn a user' recurring actions in ambient systems. ADCAIJ: Adv. Distrib. Comput. Artif. Intell. J. **1**(3), 1–10 (2013)
15. Iglesias, E.L., Borrajo, L., Romero, R.: A hmm text classification model with learning capacity. ADCAIJ: Adv. Distrib. Comput. Artif. Intell. J. **3**(3), 21 (2014)
16. Isaza, G., Mejía, M.H., Castillo, L.F., Morales, A., Duque, N.: Network management using multi-agents system. ADCAIJ: Adv. Distrib. Comput. Artif. Intell. J. **1**(3), 49–54 (2012)
17. Jasmine, K., Rajashekar, P.I., Devi, K.S., et al.: Inference in belief network using logic sampling and likelihood weighing algorithms. ADCAIJ: Adv. Distrib. Comput. Artif. Intell. J. **2**(3), 1–7 (2013)
18. López-Fernández, H., Reboiro-Jato, M., Pérez Rodríguez, J.A., Fdez-Riverola, F., Glez-Peña, D., et al.: The artificial intelligence workbench: a retrospective review (2016)
19. López Sánchez, D., Revuelta Herrero, J., de la Prieta Pintado, F., Dang, C., et al.: Analysis and visualisation of social user communities (2015)
20. Martins, C., Silva, A.R., Martins, C., Marreiros, G.: Supporting informed decision making in prevention of prostate cancer. ADCAIJ: Adv. Distrib. Comput. Artif. Intell. J. **3**(3), 1–11 (2014)
21. Mata, A., Corchado, J.M.: Forecasting the probability of finding oil slicks using a cbr system. Expert Syst. Appl. **36**(4), 8239–8246 (2009)

22. Mezquita, Y., Casado, R., Gonzalez-Briones, A., Prieto, J., Corchado, J.M.: Blockchain technology in IoT systems: review of the challenges. Ann. Emer. Technol. Comput. (AETiC) **3**(5), 17–24 (2019)
23. Mezquita, Y., Gazafroudi, A.S., Corchado, J., Shafie-Khah, M., Laaksonen, H., Kamišalić, A.: Multi-agent architecture for peer-to-peer electricity trading based on blockchain technology. In: 2019 XXVII International Conference on Information, Communication and Automation Technologies (ICAT) pp. 1–6. IEEE (2019)
24. Mezquita, Y., González-Briones, A., Casado-Vara, R., Chamoso, P., Prieto, J., Corchado, J.M.: Blockchain-based architecture: a mas proposal for efficient agri-food supply chains. In: International Symposium on Ambient Intelligence, pp. 89–96. Springer (2019)
25. Mezquita, Y., Valdeolmillos, D., González-Briones, A., Prieto, J., Corchado, J.M.: Legal aspects and emerging risks in the use of smart contracts based on blockchain. In: International Conference on Knowledge Management in Organizations, pp. 525–535. Springer (2019)
26. Moreno-Munoz, A., Bellido-Outeirino, F., Siano, P., Gomez-Nieto, M.: Mobile social media for smart grids customer engagement: emerging trends and challenges. Renew. Sustain. Energy Rev. **53**, 1611–1616 (2016)
27. Pinto, T., Marques, L., Sousa, T.M., Praça, I., Zita, V., Abreu, S.L.: Data-mining-based filtering to support solar forecasting methodologies. ADCAIJ: Adv. Distrib. Comput. Artif. Intell. J. **6**(3), 85–102 (2017)
28. Martín del Rey, A., Casado Vara, R., Hernández Serrano, D.: Reversibility of symmetric linear cellular automata with radius r = 3. Mathematics **7**(9), 816 (2019)
29. Ringler, P., Keles, D., Fichtner, W.: Agent-based modelling and simulation of smart electricity grids and markets-a literature review. Renew. Sustain. Energy Rev. **57**, 205–215 (2016)
30. Rossi, S., Barile, F.: Dominance weighted social choice functions for group recommendations. ADCAIJ: Adv. Distrib. Comput. Artif. Intell. J. **4**(1), 65–79 (2015)
31. Valdeolmillos, D., Mezquita, Y., González-Briones, A., Prieto, J., Corchado, J.M.: Blockchain technology: a review of the current challenges of cryptocurrency. In: International Congress on Blockchain and Applications, pp. 153–160. Springer (2019)
32. Valdivia, A.K.C.: Between the profiles pay per view and the protection of personal data: the product is you. ADCAIJ: Adv. Distrib. Comput. Artif. Intell. J. **6**(1), 51–58 (2017)

Data Management Applied to Service Provision in Banking Environments

Elena Hernández Nieves[✉]

BISITE Digital Innovation Hub, University of Salamanca,
Edificio Multiusos I+D+I, 37007 Salamanca, Spain
elenahn@usal.es.com

Abstract. This research aims to manage the huge amount of data that companies possess, providing a valuable tool for each of the users involved in the process. The technology derived from the use of the Big Data will allow observing and analyzing the information to assist in the decision-making process of the users of the platform who might be interested. It is intended that the Fintech technologies considered will be able to provide better efficiency and allow optimization of resources.

Keywords: Hybrid artificial intelligence system · Big Data · Machine learning · Support decision system

1 Introduction

The banking and insurance sector are immersed in a deep transformation process in which an increasing number of services are offered digitally with the aim of being more efficient and less costly. At present, these tools require an analyst or an expert to interpret the data obtained. It is to be taken into account that many times the platforms incorporate a static notification system without incorporating new visualization mechanisms that facilitate the understanding of them. Therefore, it leaves to the interpretation of an expert user the usefulness or not of the extracted data, not having an intelligence that facilitates and requires the information needed by the company. In this research the starting point is a data persistence format in which modules for knowledge management, information extraction, classification, modelling, and treatment of large volumes of data are defined and can be synthesized (Fig. 1). The information collected will also be incorporated into a reasoning system (machine learning) [1–18] that will be essential for the construction of a support tool for decision making. The objective of hybrid artificial intelligence systems, which may incorporate neural networks, CBR, Bayesian networks, fuzzy systems, etc., [19–36] is to provide to an expert information on the classification made by the system through the generation of rules that will serve as support for decision making in business environments such as banking. Hybrid systems will be essential for decision support [36–58].

J. Prieto et al. (Eds.): BLOCKCHAIN 2020, AISC 1238, pp. 165–170, 2020.
https://doi.org/10.1007/978-3-030-52535-4_19

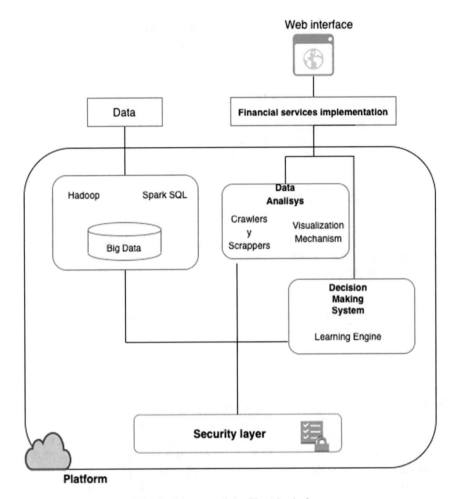

Fig. 1. Diagram of the Fintech platform

2 Conclusion

Managing and processing large volumes of data at runtime is a challenge today, as there is no single tool that offers a complete solution. The solution is to define a proposal to select a configuration of technologies to meet the needs of the system. A system of persistence is designed to ensure compatibility and homogeneity of data. Indispensable where there is a large amount of related information. The aim is to study and create mechanisms that are capable of integrating the Big Data computing model. Given the immensity of information, it is necessary to study and design Big Data and distributed computing techniques to perform an efficient data analysis and also to adjust to time constraints. In the context of the research, banking environments, it should be considered that the use of Big Data is made to treat financial data in cooperative environments, so it is of special interest both the modelling of the persistence mechanisms and the access to the data. The platform will be designed taking into account that it will implement a financial activity information analysis engine that will include the development of intelligent search, extraction and enrichment algorithms.

Acknowledgments. This research is supported by the Ministry of Education of the Junta de Castilla y León and the European Social Fund through a grant from predoctoral recruitment of research personnel associated with the University of Salamanca research project "ROBIN: Robo-advisor intelligent".

References

1. Martín del Rey, A., Batista, F.K., Queiruga Dios, A.: Malware propagation in Wireless Sensor Networks: global models vs Individual-based models. ADCAIJ: Adv. Distrib. Comput. Artif. Intell. J., Salamanca, **6**(3) (2017). ISSN: 2255-2863
2. Baruque, B., Corchado, E., Mata, A., Corchado, J.M.: A forecasting solution to the oil spill problem based on a hybrid intelligent system. Inf. Sci. **180**(10), 2029–2043 (2010)
3. Durik, B.O.: Organisational metamodel for large-scale multi-agent systems: first steps towards modelling organisation dynamics. ADCAIJ: Adv. Distrib. Comput. Artif. Intell. J. Salamanca, **6**(3) (2017). ISSN: 2255-2863
4. Bullon, J., et al.: Manufacturing processes in the textile industry. Expert Systems for fabrics production. ADCAIJ: Adv. Distrib. Comput. Artif. Intell. J. **6**(4), 15–23 (2017)
5. González, C., Burguillo, J.C., Llamas, M., Laza, R.: Designing intelligent tutoring systems: a personalization strategy using case-based reasoning and multi-agent systems. ADCAIJ: Advances in Distributed Computing and Artificial Intelligence Journal, Salamanca, vol. 2, no. 1 (2013). ISSN: 2255-2863
6. Casado-Vara, R., Chamoso, P., De la Prieta, F., Prieto, J., Corchado, J.M.: Non-linear adaptive closed-loop control system for improved efficiency in IoT-blockchain management. Inf. Fusion **49**, 227–239 (2019)
7. Casado-Vara, R., Martin-del Rey, A., Affes, S., Prieto, J., Corchado, J.M.: IoT network slicing on virtual layers of homogeneous data for improved algorithm operation in smart buildings. Fut. Generation Comput. Syst. **102**, 965–977 (2020)
8. Casado-Vara, R., Novais, P., Gil, A.B., Prieto, J., Corchado, J.M.: Distributed continuous-time fault estimation control for multiple devices in IoT networks. IEEE Access **7**, 11972–11984 (2019)
9. Casado-Vara, R., Prieto, J., De la Prieta, F., Corchado, J.M.: How blockchain improves the supply chain: case study alimentary supply chain. Procedia Comput. Sci. **134**, 393–398 (2018)
10. Chamoso, P., González-Briones, A., Rodríguez, S., Corchado, J.M.: Tendencies of technologies and platforms in smart cities: a state-of-the-art review. Wireless Commun. Mob. Comput. **2018**, 1–17 (2018)
11. Choon, Y.W., Mohamad, M.S., Safaai Deris, R.M., Illias, C.K.C., Chai, L.E., Omatu, S., Corchado, J.M.: Differential bees flux balance analysis with OptKnock for in silico microbial strains optimization. PloS One, **9**(7), e102744 (2014)
12. Corchado, J. M., Aiken, J.: Hybrid artificial intelligence methods in oceanographic forecast models. IEEE Trans. Syst. Man Cybern. Part C (Appl. Rev.), **32**(4), 307–313 (2002)
13. Corchado, J.M., Fyfe, C.: Unsupervised neural method for temperature forecasting. Artif. Intell. Eng. **13**(4), 351–357 (1999)

14. Corchado, J.M., Lees, B.: A hybrid case-based model for forecasting. Appl. Artif. Intell. **15** (2), 105–127 (2001)
15. Corchado, J.M., Corchado, E.S., Aiken, J., Fyfe, C., Fernandez, F., Gonzalez, M.: Maximum likelihood hebbian learning based retrieval method for CBR systems. In: International Conference on Case-Based Reasoning, pp. 107–121. Springer, Heidelberg (2003)
16. Corchado, J.M., Pavón, J., Corchado, E.S., Castillo, L.F.: Development of CBR-BDI agents: a tourist guide application. In: European Conference on Case-based Reasoning, pp. 547–559. Springer, Heidelberg (2004)
17. Coria, J.A.G., Castellanos-Garzón, J.A., Corchado, J.M.: Intelligent business processes composition based on multi-agent systems. Expert Syst. Appl. **41**(4), 1189–1205 (2014)
18. Cunha, R., Cleo, B., Diana, A.: Development of a graphical tool to integrate the Prometheus AEOlus methodology and Jason Platform. ADCAIJ: Adv. Distrib. Comput. Artif. Intell. J., **6** (2), 57–70 (2017)
19. Daniel, A., Roldán, J.C., Ruiz, D., Gallego, F.O.: An approach for discovering keywords from Spanish tweets using Wikipedia. ADCAIJ: Adv. Distrib. Comput. Artificial Intell. J., Salamanca, **4**(2) (2015). ISSN: 2255-2863
20. Díaz, F., Fdez-Riverola, F., Corchado, J.M.: gene-CBR: a case-based reasonig tool for cancer diagnosis using microarray data sets. Comput. Intell. **22**(3–4), 254–268 (2006)
21. Farias, G.P., et al.: Predicting plan failure by monitoring action sequences and duration. ADCAIJ: Adv. Distrib. Comput. Artif. Intell. J. **6**(4), 55–69 (2017)
22. Fdez-Riverola, F., Corchado, J.M.: Fsfrt: forecasting system for red tides. Appl. Intell. **21**(3), 251–264 (2004)
23. Fdez-Riverola, F., Iglesias, E.L., Díaz, F., Méndez, J.R., Corchado, J.M.: Applying lazy learning algorithms to tackle concept drift in spam filtering. Expert Syst. Appl. **33**(1), 36–48 (2007)
24. Fdez-Riverola, F., Iglesias, E.L., Díaz, F., Méndez, J.R., Corchado, J.M.: SpamHunting: An instance-based reasoning system for spam labelling and filtering. Decision Supp. Syst. **43**(3), 722–736 (2007)
25. Fernández-Riverola, F., Diaz, F., Corchado, J.M.: Reducing the memory size of a fuzzy case-based reasoning system applying rough set techniques. IEEE Trans. Syst. Man Cybern. Part C (Applications and Reviews), **37**(1), 138–146 (2006)
26. Fyfe, C., Corchado, J.M.: Automating the construction of CBR systems using kernel methods. Int. J. Intell. Syst. **16**(4), 571–586 (2001)
27. Giovani, P.F., Ramon, F.P., Lucas, W.H., Felipe, M., Renata, V., Rafael, H.B.: Predicting plan failure by monitoring action sequences and duration. ADCAIJ: Adv. Distrib. Comput. Artif. Intell. J., Salamanca, **6**(2) (2017). ISSN: 2255-2863
28. Stefania da Silveira, G., et al.: Modeling of circadian rhythm under influence of pain: an approach based on multi-agent simulation. ADCAIJ: Adv. Distrib. Comput. Artif. Intell. J. **7** (2), 17–25 (2018)
29. Glez-Bedia, M., Corchado, J.M., Corchado, E.S., Fyfe, C.: Analytical model for constructing deliberative agents. Eng. Intell. Syst. Electric. Eng. Commun. **10**(3), 173–185 (2002)
30. González-Briones, A., Prieto, J., De La Prieta, F., Herrera-Viedma, E., Corchado, J.M.: Energy optimization using a case-based reasoning strategy. Sensors **18**(3), 865 (2018)
31. Guillén, J.H., del Rey, A.M., Casado-Vara, R.: Security Countermeasures of a SCIRAS model for advanced malware propagation. IEEE Access **7**, 135472–135478 (2019)
32. Javier, G., Xavier, A., Germán, M., Juan, C.T., Adalberto, P.: AmICog – mobile technologies to assist people with cognitive disabilities in the workplace. DCAIJ: Adv. Distrib. Comput. Artif. Intell. J., Salamanca, **2**(4) (2013). ISSN: 2255-2863

33. Garzón, J.A.C., González, J.R.: A gene selection approach based on clustering for classification tasks in colon cancer. ADCAIJ: Adv. Distrib. Comput. Artif. Intell. J., Salamanca, **4**(3) (2015). ISSN: 2255-2863

34. Kethareswaran, V., Sankar Ram, C.: An indian perspective on the adverse impact of internet of things (IoT). ADCAIJ: Adv. Distrib. Comput. Artif. Intell. J. **6**(4), 35–40 (2017)

35. Li, T., Sun, S., Bolić, M., Corchado, J.M.: Algorithm design for parallel implementation of the SMC-PHD filter. Signal Process. **119**, 115–127 (2016)

36. Li, T., Sun, S., Corchado, J. M., Siyau, M.F.: A particle dyeing approach for track continuity for the SMC-PHD filter. In: 17th International Conference on Information Fusion (FUSION), pp. 1–8. IEEE, July 2014

37. Lima, A.C.E., de Castro, L.N., Corchado, J.M.: A polarity analysis framework for Twitter messages. Appl. Math. Comput. **270**, 756–767 (2015)

38. Mar, L., Juanita, P., Javier, C., Molina, J.M.: The awareness of Privacy issues in Ambient Intelligence. ADCAIJ: Adv. Distrib. Comput. Artif. Intell. J., Salamanca, **3**(2) (2014). (ISSN: 2255–2863)

39. Martín del Rey, A., Casado Vara, R., Hernández Serrano, D.: Reversibility of symmetric linear cellular automata with radius r = 3. Mathematics **7**(9), 816 (2019)

40. Mata, A., Corchado, J.M.: Forecasting the probability of finding oil slicks using a CBR system. Expert Syst. Appl. **36**(4), 8239–8246 (2009)

41. Mendez, J.R., Fdez-Riverola, F., Diaz, F., Iglesias, E.L., Corchado, J.M.: A comparative performance study of feature selection methods for the anti-spam filtering domain. In: Industrial Conference on Data Mining, pp. 106–120. Springer, Heidelberg (2006)

42. Miki, U., Naoki, M., Keinosuke, M.: Picture models for 2-scene comics creating system. ADCAIJ: Adv. Distrib. Comput. Artif. Intell. J., Salamanca, **3**(2) (2014). ISSN: 2255-2863

43. Ming, F.S., Li, T., Loo, J.: A novel pilot expansion approach for MIMO channel estimation. ADCAIJ: Adv. Distrib. Comput. Artif. Intell. J. Salamanca, **3**(3) (2014). ISSN: 2255-2863

44. Morente-Molinera, J.A., Kou, G., González-Crespo, R., Corchado, J.M., Herrera-Viedma, E.: Solving multi-criteria group decision making problems under environments with a high number of alternatives using fuzzy ontologies and multi-granular linguistic modelling methods. Knowl. Based Syst. **137**, 54–64 (2017)

45. Moung, E.: A comparison of the YCBCR color space with gray scale for face recognition for surveillance applications. ADCAIJ: Adv. Distrib. Comput. Artif. Intell. J. **6**(4), 25–33 (2017)

46. Pawel, P., Kamila, K.: Modeling and simulation of bus assembling process using DES/ABS approach. ADCAIJ: Adv. Distrib. Comput. Artif. Intell. J., Salamanca, **6**(1) (2017). ISSN: 2255-2863

47. Ribeiro, C., et al.: Customized normalization clustering meth-odology for consumers with heterogeneous characteristics. ADCAIJ: Adv. Distrib. Comput. Artif. Intell. J. **7**(2), 53–69 (2018)

48. Ricardo, A.S., Rafaela, L.C., Ronaldo, L.R.C., Jonas, V., Fernando De La, P.: Learning objects recommendation system: issues and approaches for retrieving, indexing and recomend learning objects. ADCAIJ: Adv. Distrib. Comput. Artif. Intell. J., Salamanca, **4**(4) (2015). ISSN: 2255-2863

49. Ricardo, S., Da Silva Bitencourt, G.K., Gelaim, T.A., Marchi, J., De La Prieta, F.: Towards a model of open and reliable cognitive multiagent systems: dealing with trust and emotions. ADCAIJ: Adv. Distrib. Comput. Artif. Intell. J., Salamanca, **4**(3) (2015). ISSN: 2255-2863

50. Saadi, B.A.K., Md Ghanib, N.A., Liong, C-Y., Jemain, A.A.: Firearm classification using neural networks on ring of firing pin impression images. ADCAIJ: Adv. Distrib. Comput. Artif. Intell. J., Salamanca, **1**(3) (2012). ISSN: 2255-2863

51. Serna, F.J.A., Iniesta, J.B.: The delimitation of freedom of speech on the Internet: the confrontation of rights and digital censorship. ADCAIJ: Adv. Distrib. Comput. Artificial Intell. J. **7**(1), 5–12 (2018)
52. de Castro, S., Fernando, L., Alves, G.V., Borges, A.P.: Using trust degree for agents in order to assign spots in a Smart Parking (2017)
53. Srivastava, V., Ravindra, P.: An extension of local mesh peak valley edge based feature descriptor for image retrieval in bio-medical images. ADCAIJ: Adv. Distrib. Comput. Artif. Intell. J. **7**(1), 77–89 (2018)
54. Tapia, D.I., Corchado, J.M.: An ambient intelligence based multi-agent system for alzheimer health care. Int. J. Ambient Comput. Intell. (IJACI) **1**(1), 15–26 (2009)
55. Tapia, D.I., Fraile, J.A., Rodríguez, S., Alonso, R.S., Corchado, J.M.: Integrating hardware agents into an enhanced multi-agent architecture for Ambient Intelligence systems. Inf. Sci. **222**, 47–65 (2013)
56. Heijmeijer, V.H., Alexis, Alves, G.V.: Development of a middleware between SUMO simulation tool and JaCaMo framework. ADCAIJ: Adv. Distrib. Comput. Artif. Intell. J. **7** (2), 5–15 (2018)
57. Vanessa, N.C., Hisham, M.H., Hossain, S.: Android malware detection using kullback-leibler divergence. ADCAIJ: Adv. Distrib.Comput. Artif. Intell. J., Salamanca, **3**(2) (2014). ISSN: 2255–2863
58. Vera, J., Stewart, E.: Human rights in the ethical protection of youth in social networks-the case of Colombia and Peru. ADCAIJ: Adv. Distrib. Comput. Artif. Intell. J. **6**(4), 71–79 (2017)

An Intelligent Platform for the Monitoring and Evaluation of Critical Marine Infrastructures

Marta Plaza-Hernández[✉]

BISITE Research Group, University of Salamanca, Edificio Multiusos I+D+i,
Calle Espejo 2, 37007 Salamanca, Spain
martaplaza@usal.es

Abstract. Over the past five years the Internet of Things (IoT) technology has grown rapidly, finding applications in several sectors. Large shipping industries are already investing in IoT techniques to optimise transparency, safety and reduce costs. This research proposal aims to develop a smart platform that facilitates the management of critical marine systems, contributing to the generation and transfer of knowledge in the fields of Edge Computing, Intelligent Models and Virtual Organisations.

Keywords: Internet of Things · Artificial Intelligence · Edge Computing · Smart Maritime Industry

1 Introduction

The Internet of Things (IoT) is a network of physical "smart" devices embedded with electronics, software, sensors and actuators, that allows interconnectivity among devices and data exchange. This new technology has grown rapidly [1], finding applications in several sectors [2] (e.g. energy, healthcare, industrial, IT and networks, security and public safety and transportation). In the maritime industry, the implementation of the IoT technology enables shipping companies to connect their vessels in one platform, allowing data sharing with the entire corporate ecosystem that stakeholders can exploit for decision-making. IoT systems are set to:

- improve the efficiency of the sector's activities
- improve the transparency of companies and institutions
- increase the safety and well-being of workers on-board
- reduce inefficiencies, risks and costs
- minimise the environmental impact

For the transportation and logistics sectors, which have always relied on exchanging decision-making data, the digitalisation process has been easy, placing them ahead in the transition. However, the maritime industry, heavily anchored in traditional methodologies, is facing several obstacles mainly because it operates in some of the most remote areas of the planet, where M2M interactions are complicated.

© The Editor(s) (if applicable) and The Author(s), under exclusive license
to Springer Nature Switzerland AG 2020
J. Prieto et al. (Eds.): BLOCKCHAIN 2020, AISC 1238, pp. 171–176, 2020.
https://doi.org/10.1007/978-3-030-52535-4_20

Many institutions from the public and private sectors are making great efforts to facilitate the transition towards the digitalisation of the maritime sector. The European Union, through its Horizon 2020 programme, will allocate up to EUR 6.3 billion for research and development of ICT and IoT technologies [3, 4]. It is expected that by 2025, IoT will reach a potential market impact of USD 11.1 trillion [5].

This research proposal aims to develop an intelligent platform that allows the inclusion of Artificial Intelligence (AI) algorithms and models [6–17] for the management of critical marine systems. It will have the capacity to combine information stored in databases with data acquired in real-time [18–37]. To solve critical systems, we will design an architecture that facilitates the integration of intelligent algorithms capable of managing data and information in real-time, responding in execution time, and that have backup mechanisms. Also, we will attempt to introduce new algorithms based on automatic learning techniques that help to create intelligent systems [38–47]. The platform will use the power of a cloud for decision-making and the flexibility to distribute intelligence to the edge of the network [48–64].

The "Surveying & MARiTime internet of thingS EducAtion (SMARTSEA)" is a project funded by the EU Erasmus+ Programme [65]. It aims to develop an interactive MSc course on Maritime and Surveyor ICT/IoT systems, helping complete the market void in technical and maintenance specialists generated by the prompt expansion of the Smart Maritime & Surveying industry. The intelligent platform proposed here will facilitate the unification of all the technology developed by the SMARTSEA project consortium.

2   Conclusions

IoT is considered one of the leading gateway technologies to digital transformation. This work aims to develop an intelligent platform for the management of critical marine systems, within the framework of the SMARTSEA project.

This research work will analyse intelligent systems that allow decentralised decision-making, reducing network latency, facilitating the process in case of communication failures and increasing security. The platform will have greater autonomy and scalability, facilitating its integration with other systems. Moreover, this work will promote the interaction between naval organisations and port environments.

Acknowledgments. This research has been supported by the project "The Surveying & MARiTime internet of thingS EducAtion (SMARTSEA)", Reference: 612198-EPP-1-2019-1-ES-EPPKA2-KA, financed by the European Commission (Erasmus+ : Higher Education - International Capacity Building).

References

1. Manyika, J., Chui, M., Bisson, P., Woetzel, J., Dobbs, R., Bughin, J., Aharon, D.: The internet of things: mapping the value beyond the hype. McKinsey Global Institute (2015)

2. Beecham Research Homepage. M2M Sector Map. http://beechamresearch.com/. Accessed 12 Jan 2020
3. European Commission. EU leads the way with ambitious action for cleaner and safer seas. https://ourocean2017.org/eu-leads-way-ambitious-action-cleaner-and-safer-seas. Accessed 07 Jan 2020
4. European Commission. Horizon 2020 - smart, green and integrated transport. https://ec.europa.eu/programmes/horizon2020/en/h2020-section/smart-green-and-integrated-transport. Accessed 07 Jan 2020
5. Deloitte. https://www2.deloitte.com/tr/en/pages/technology-media-and-telecommunications/articles/internet-of-things-iot-in-shipping-industry.html. Accessed 09 Jan 2020
6. Li, T., Sun, S., Corchado, J. M., Siyau, M.F.: A particle dyeing approach for track continuity for the SMC-PHD filter. In: 17th International Conference on Information Fusion (FUSION), pp. 1–8. IEEE, July 2014
7. Wang, X., Tarrío, P., Bernardos, A.M., Metola, E., Casar, J.R.: User-independent accelerometer based-gesture recognition for mobile devices. ADCAIJ: Adv. Distrib. Comput. Artif. Intell. J. $\mathbf{1}$(3), 11–25 (2012). ISSN: 2255-2863
8. Urbano, J., Cardoso, H.L., Rocha, A.P., Oliveira, E.: Trust and normative control in multi-agent systems. ADCAIJ: Adv. Distrib. Comput. Artif. Intell. J. $\mathbf{1}$(1) (2012). (ISSN: 2255-2863)
9. Oliveira, T., Neves, J., Novais, P.: Guideline formalization and knowledge representation for clinical decision support. ADCAIJ: Adv. Distrib. Comput. Artif. Intell. J. $\mathbf{1}$(2), 1–11 (2012). ISSN: 2255-2863
10. Fdez-Riverola, F., Iglesias, E.L., Díaz, F., Méndez, J.R., Corchado, J.M.: Applying lazy learning algorithms to tackle concept drift in spam filtering. Expert Syst. Appl. $\mathbf{33}$(1), 36–48 (2007)
11. Aige, M.B.: The online tourist fraud: the new measures of technological investigation in Spain. ADCAIJ: Adv. Distrib. Comput. Artif. Intell. J. $\mathbf{6}$(2), 85 (2017). ISSN: 2255-2863
12. Morente-Molinera, J.A., Kou, G., González-Crespo, R., Corchado, J.M., Herrera-Viedma, E.: Solving multi-criteria group decision making problems under environments with a high number of alternatives using fuzzy ontologies and multi-granular linguistic modelling methods. Knowl.-Based Syst. $\mathbf{137}$, 54–64 (2017)
13. Carneiro, D., Araújo, D., Pimenta, A., Novais, P.: Real time analytics for characterizing the computer user's state. ADCAIJ: Adv. Distrib. Comput. Artif. Intell. J. $\mathbf{5}$(4) (2016). ISSN: 2255-2863
14. Li, T., Sun, S., Bolić, M., Corchado, J.M.: Algorithm design for parallel implementation of the SMC-PHD filter. Signal Process. $\mathbf{119}$, 115–127 (2016)
15. Coria, J.A.G., Castellanos-Garzón, J.A., Corchado, J.M.: Intelligent business processes composition based on multi-agent systems. Expert Syst. Appl. $\mathbf{41}$(4), 1189–1205 (2014)
16. Silva, A., Oliveira, T., Neves, J., Novais, P.: Treating colon cancer survivability prediction as a classification problem. ADCAIJ: Adv. Distrib. Comput. Artif. Intell. J. $\mathbf{5}$(1), 37 (2016). ISSN: 2255-2863
17. Tapia, D.I., Fraile, J.A., Rodríguez, S., Alonso, R.S., Corchado, J.M.: Integrating hardware agents into an enhanced multi-agent architecture for Ambient Intelligence systems. Inf. Sci. $\mathbf{222}$, 47–65 (2013)
18. Corchado, J.M., Pavón, J., Corchado, E.S., Castillo, L.F. Development of CBR-BDI agents: a tourist guide application. In: European Conference on Case-based Reasoning, pp. 547–559. Springer, Berlin, August 2004
19. Lima, A.C.E., de Castro, L.N., Corchado, J.M.: A polarity analysis framework for Twitter messages. Appl. Math. Comput. $\mathbf{270}$, 756–767 (2015)

20. Nihan, C.E.: Healthier? More efficient? Fairer? An overview of the main ethical issues raised by the use of ubicomp in the workplace. ADCAIJ: Adv. Distrib. Comput. Artif. Intell. J. **2** (1), 29–40 (2013). ISSN: 2255-2863

21. Macek, K., Rojicek, J., Kontes, G., Rovas, D.V.: Black-box optimization for buildings and its enhancement by advanced communication infrastructure. ADCAIJ: Adv. Distrib. Comput. Artif. Intell. J. **1**(5), 53–64 (2013). ISSN: 2255-2863

22. Fdez-Riverola, F., Corchado, J.M.: FSfRT: forecasting system for red tides. Appl. Intell. **21** (3), 251–264 (2004)

23. Fdez-Riverola, F., Iglesias, E.L., Díaz, F., Méndez, J.R., Corchado, J.M.: SpamHunting: an instance-based reasoning system for spam labelling and filtering. Decis. Support Syst. **43**(3), 722–736 (2007)

24. Casado-Vara, R., Martin-del Rey, A., Affes, S., Prieto, J., Corchado, J.M.: IoT network slicing on virtual layers of homogeneous data for improved algorithm operation in smart buildings. Future Gener. Comput. Syst. **102**, 965–977 (2020)

25. Ueno, M., Suenaga, T., Isahara, H.: Classification of two comic books based on convolutional neural networks. ADCAIJ: Adv. Distrib. Comput. Artif. Intell. J. **6**(1), 5 (2017). ISSN: 2255-2863

26. Baruque, B., Corchado, E., Mata, A., Corchado, J.M.: A forecasting solution to the oil spill problem based on a hybrid intelligent system. Inf. Sci. **180**(10), 2029–2043 (2010)

27. Casado-Vara, R., Prieto, J., De la Prieta, F., Corchado, J.M.: How blockchain improves the supply chain: case study alimentary supply chain. Procedia Comput. Sci. **134**, 393–398 (2018)

28. Silva, F., Analide, C.: Tracking context-aware well-being through intelligent environments. ADCAIJ: Adv. Distrib. Comput. Artif. Intell. J. **4**(2), 61 (2015). ISSN: 2255-2863

29. Li, T., Sun, S.: Online adapting the magnitude of target birth intensity in the PHD filter. ADCAIJ: Adv. Distrib. Comput. Artif. Intell. J **2**(4), 31 (2013). ISSN: 2255-2863

30. Corchado, J.M., Aiken, J.: Hybrid artificial intelligence methods in oceanographic forecast models. IEEE Trans. Syst. Man Cybern. Part C (Appl. Rev.) **32**(4), 307–313 (2002)

31. González-Briones, A., Prieto, J., De La Prieta, F., Herrera-Viedma, E., Corchado, J.M.: Energy optimization using a case-based reasoning strategy. Sensors **18**(3), 865 (2018)

32. Díaz, F., Fdez-Riverola, F., Corchado, J.M.: gene-CBR: a case-based reasoning tool for cancer diagnosis using microarray data sets. Comput. Intell. **22**(3–4), 254–268 (2006)

33. Corchado, J.M., Corchado, E.S., Aiken, J., Fyfe, C., Fernandez, F., Gonzalez, M.: Maximum likelihood hebbian learning based retrieval method for cbr systems. In: International Conference on Case-Based Reasoning, pp. 107–121. Springer, Berlin, June 2003

34. Martinez-Martin, E., Escrig, M.T., Del Pobil, A.P.: A qualitative acceleration model based on intervals. ADCAIJ: Adv. Distrib. Comput. Artif. Intell. J. **2**(2) (2013). ISSN: 2255-2863

35. Guillén, J.H., del Rey, A.M., Casado-Vara, R.: Security countermeasures of a SCIRAS model for advanced malware propagation. IEEE Access **7**, 135472–135478 (2019)

36. Corchado, J.M., Lees, B.: A hybrid case-based model for forecasting. Appl. Artif. Intell. **15** (2), 105–127 (2001)

37. Satoh, I.: Bio-inspired self-adaptive agents in distributed systems. ADCAIJ: Adv. Distrib. Comput. Artif. Intell. J. **1**(2) (2012). ISSN: 2255-2863

38. Fernández-Riverola, F., Diaz, F., Corchado, J.M.: Reducing the memory size of a fuzzy case-based reasoning system applying rough set techniques. IEEE Trans. Syst. Man Cybern. Part C (Appl. Rev.) **37**(1), 138–146 (2006)

39. Tapia, D.I., Corchado, J.M.: An ambient intelligence based multi-agent system for alzheimer health care. Int. J. Ambient Comput. Intell. (IJACI) **1**(1), 15–26 (2009)

40. Adam, E., Grislin-Le Strugeon, E., Mandiau, R.: MAS architecture and knowledge model for vehicles data communication. ADCAIJ: Adv. Distrib. Comput. Artif. Intell. J. **1**(1), 23–31 (2012). ISSN: 2255-2863

41. Corchado, J.M., Fyfe, C.: Unsupervised neural method for temperature forecasting. Artif. Intell. Eng. **13**(4), 351–357 (1999)

42. Mendez, J.R., Fdez-Riverola, F., Diaz, F., Iglesias, E.L., Corchado, J.M. A comparative performance study of feature selection methods for the anti-spam filtering domain. In: Industrial Conference on Data Mining, pp. 106–120. Springer, Berlin, July 2006

43. Mata, A., Corchado, J.M.: Forecasting the probability of finding oil slicks using a CBR system. Expert Syst. Appl. **36**(4), 8239–8246 (2009)

44. Chamoso, P., González-Briones, A., Rodríguez, S., Corchado, J.M.: Tendencies of technologies and platforms in smart cities: a state-of-the-art review. Wirel. Commun. Mobile Comput. **2018**, 1–17 (2018)

45. Glez-Bedia, M., Corchado, J.M., Corchado, E.S., Fyfe, C.: Analytical model for constructing deliberative agents. Eng. Intell. Syst. Electr. Eng. Commun. **10**(3), 173–185 (2002)

46. Ochoa-Aday, L., Cervelló-Pastor, C., Fernández-Fernández, A.: Discovering the network topology: an efficient approach for SDN. ADCAIJ: Adv. Distrib. Comput. Artif. Intell. J **5**(2), 101 (2016). ISSN: 2255-2863

47. Fyfe, C., Corchado, J.M.: Automating the construction of CBR Systems using Kernel Methods. Int. J. Intell. Syst. **16**(4), 571–586 (2001)

48. Choon, Y.W., et al.: Differential bees flux balance analysis with OptKnock for in silico microbial strains optimization. PLoS ONE **9**(7), e102744 (2014)

49. Li, T., Sun, S., Corchado, J.M., Siyau, M.F.: A particle dyeing approach for track continuity for the SMC-PHD filter. In: 17th International Conference on Information Fusion (FUSION), pp. 1–8. IEEE, July 2014

50. Pawlewski, P., Golinska, P., Dossou, P.-E.: Application potential of agent based simulation and discrete event simulation in enterprise integration modelling concepts. ADCAIJ: Adv. Distrib. Comput. Artif. Intell. J. **1**(1), 33–42 (2012). ISSN: 2255-2863

51. Martín del Rey, A., Casado Vara, R., Hernández Serrano, D.: Reversibility of symmetric linear cellular automata with radius r = 3. Mathematics **7**(9), 816 (2019)

52. Ueno, M., Mori, N., Matsumoto, K.: Picture information shared conversation agent: Pictgent. ADCAIJ: Adv. Distrib. Comput. Artif. Intell. J. **1**(1) (2012) ISSN: 2255-2863

53. Griol, D., García-Herrero, J., Molina, J.M.: Combining heterogeneous inputs for the development of adaptive and multimodal interaction systems. ADCAIJ: Adv. Distrib. Comput. Artif. Intell. J. **2**(3) (2013) ISSN: 2255-2863

54. Casado-Vara, R., Novais, P., Gil, A.B., Prieto, J., Corchado, J.M.: Distributed continuous-time fault estimation control for multiple devices in IoT networks. IEEE Access **7**, 11972–11984 (2019)

55. Vilaro, A., Orero, P.: User-centric cognitive assessment. Evaluation of attention in special working centres: from paper to Kinect. DCAIJ: Adv. Distrib. Comput. Artif. Intell. J. **2**(4), 19–22 (2013). ISSN: 2255-2863

56. Romero, S., Fardoun, H.M., Penichet, V.M.R., Gallud, J.A.: Tweacher: new proposal for online social networks impact in secondary education. ADCAIJ: Adv. Distrib. Comput. Artif. Intell. J. **2**(1), 9–18 (2013). ISSN: 2255-2863

57. Fuentes, D., Laza, R., Pereira, A.: Intelligent devices in rural wireless networks. DCAIJ: Adv. Distrib. Comput. Artif. Intell. J. **2**(4), 23–30 (2013). ISSN: 2255-2863

58. Macintosh, A., Feisiyau, M., Ghavami, M.: Impact of the mobility models, route and link connectivity on the performance of position based routing protocols. ADCAIJ: Adv. Distrib. Comput. Artif. Intell. J. **3**(1), 74–91 (2014). ISSN: 2255-2863

59. Casado-Vara, R., Chamoso, P., De la Prieta, F., Prieto, J., Corchado, J.M.: Non-linear adaptive closed-loop control system for improved efficiency in IoT-blockchain management. Inf. Fusion **49**, 227–239 (2019)
60. Alam, N., Sultana, M., Alam, M.S., Al-Mamun, M.A., Hossain, M.A.: Optimal intermittent dose schedules for chemotherapy using genetic algorithm. ADCAIJ: Adv. Distrib. Comput. Artif. Intell. J. **2**(2), 37–52 (2013). ISSN: 2255-2863
61. Magaña, V.C., Organero, M.M., Álvarez-García, J.A., Rodríguez, J.Y.F.: Design of a speed assistant to minimize the driver stress. ADCAIJ: Adv. Distrib. Comput. Artif. Intell. J. **6**(3), 45 (2017). ISSN: 2255-2863
62. Marín, P.A.R., Giraldo, M., Tabares, V., Duque, N., Ovalle, D.: Educational resources recommendation system for a heterogeneous student group. ADCAIJ: Adv. Distrib. Comput. Artif. Intell. J. **5**(3), 21–30 (2016). ISSN: 2255-2863
63. Desquesnes, G., Lozenguez, G., Doniec, A., Duviella, É.: Planning large systems with MDPs: case study of inland waterways supervision. ADCAIJ: Adv. Distrib. Comput. Artif. Intell. J. **5**(4), 71 84 (2016). ISSN: 2255-2863
64. Oliver, M., Molina, J.P., Fernández-Caballero, A., González, P.: Collaborative computer-assisted cognitive rehabilitation system. ADCAIJ: Adv. Distrib. Comput. Artif. Intell. J. **6** (3), 57 (2017). ISSN: 2255-2863
65. SMARTSEA. https://www.smart-sea.eu/. Accessed 01 Mar 2020

Author Index

© The Editor(s) (if applicable) and The Author(s), under exclusive license
to Springer Nature Switzerland AG 2020
J. Prieto et al. (Eds.): BLOCKCHAIN 2020, AISC 1238, pp. 177–178, 2020.
https://doi.org/10.1007/978-3-030-52535-4

Printed in the United States
By Bookmasters